The Zygote of Christ
&
The Mystery of Man

Elizabeth Bothamley Rex

and

Francis Etheredge

En Route Books and Media, LLC
Saint Louis, MO

En Route Books and Media, LLC
5705 Rhodes Avenue
St. Louis, MO 63109

Cover Art "Zygote of Christ" is by nationally recognized artist, Nellie Edwards (nellie@paintedfaith.net), known most for her depiction of Our Lady of Guadalupe, kneeling in adoration of her full-term Savior-Son, titled "Mother of Life." See her gallery and the story behind the artwork at www.PaintedFaith.Net

Copyright © 2025 Elizabeth Bothamley Rex and Francis Etheredge

ISBN-13: 979-8-88870-378-6
Library of Congress Control Number: 2025939677

No part of this book may be reproduced, stored in a retrieval system, or transmitted in any form, or by any means, electronic, mechanical, photocopying, or otherwise, without the prior written permission of the author.

Psalm 8

How great is your name, O Lord our God,
Through all the earth!

Your majesty is praised above the heavens;
on the lips of children and of babes
you have found praise to foil your enemy,
to silence the foe and the rebel.

When I see the heavens, the work of your hands,
the moon and the stars which you arranged,
what is man that you should keep him in mind,
mortal man that you care for him?

Yet you have made him little less than a god;
with glory and honor you crowned him,
gave him power over the works of your hand,
put all things under his feet.

All of them sheep and cattle,
yes, even the savage beasts,
birds of the air, and fish,
that make their way through the waters.

How great is your name, O Lord our God,
through all the earth!

CONTENTS

Introduction

The Zygote of Christ and the Mystery of Man
Elizabeth Bothamley Rex
1

Essay I

The Human Person: A Bioethical Word

The Human Person: An Indivisibly Psychosomatic Being from Conception
Francis Etheredge
5

Essay II

The Human Person: A Bioethical Word

An Introduction to Genesis I-II: The Creation Gift of Participation in the Mystery of the Blessed Trinity
Elizabeth Bothamley Rex
25

Essay III

Conception: A Icon of the Beginning

**The Gift and Rights of Being Conceived In-Relationship:
We Are an Icon of the Beginning**
Francis Etheredge
39

Essay IV

Conception: An Icon of the Beginning

An End Word: A New Beginning
Elizabeth Bothamley Rex
71

Essay V

Journal of Bioethics in Law & Culture
Summer 2024

**The Importance of the Alabama Supreme Court Decision:
Frozen Embryos are "Extrauterine" Unborn Children**
Elizabeth Bothamley Rex
117

Essay VI

Sebastian's Point / Society of St. Sebastian
Weekly Column - 09 September 2024

Model Laws in Germany, Italy, Louisiana, and Georgia Restrict Both IVF and Cryopreservation to Legally Protect Human Embryos from Further Abuse, Physical Harm, and Death
Elizabeth Bothamley Rex
137

Essay VII

Journal of Bioethics in Law & Culture
Spring 2025

The Principle of Fraternal Charity, Organ Donation and Embryo Adoption: From Magisterial Condemnation to Magisterial Commendation
Elizabeth Bothamley Rex
137

Front Cover Art Commentary
The Zygote of Christ
Nellie Edwards
167

Introduction
Elizabeth Bothamley Rex

The Zygote of Christ and the Mystery of Man

"The truth is that only in the mystery of the incarnate Word does the mystery of man take on light."
~ Gaudium et spes, n. 22

The Church has taught us that the Mystery of Man can only be found in the Mystery of the Incarnate Word, so what better place to begin searching for the truth about the Mystery of Man than to prayerfully ponder and contemplate the "Word made flesh" – *Verbum caro factum est* - at the very first moment of the Incarnation of Christ? When and how was the Second Person of the Blessed Trinity conceived by the Holy Spirit and "made man" – *et homo factus est*? What did Jesus look like at the first moment of the Mystery of the Incarnation within Mary's womb? Was Christ actually a one cell *zygote* and a pre-implantation *embryo*, just like every one of us?

The front cover of this book features the beautiful and awe-inspiring painting of the *Zygote of Christ* by Nellie Edwards that magnificently captures and communicates the Mystery of the Trinity, the Mystery of Christ, and the Mystery of Man. Look carefully at the *Zygote of Christ* and you will even discover the "hidden" faces of both Jesus and Mary!

In the book, *The Scandal of the Incarnation*, St. Irenaeus, an early Church Father, teaches:

"The truth of this was shown when the Word of God became man, assimilating Himself to man and man to Himself, so that by His resemblance to the Son, man might become precious to the Father." (Irenaeus *Against the Heresies*, V 16,2)

In another quote, St. Irenaeus continues to defend Christ's perfect human nature by teaching:

"How could we be partakers of adoption as God's sons without receiving from Him through the Son, the gift of communion with Him? ... This is why *He passed through all the ages of human life*, restoring to all men communion with God. (Irenaeus *Against the Heresies*, II 18, 7, with emphasis added.)

The primary purpose of our book is shed greater light on the great Mystery of Man through "selected essays" taken from two excellent books written by the English author, Francis Etheredge: *The Human Person: A Bioethical Word* and *Conception: An Icon of the Beginning*. These scholarly works thoroughly present and advance one of the most important - and still unanswered - questions regarding the Mystery of Man, namely, the "timing" of God's immediate creation and infusion of our spiritual soul into the human body at the moment of conception, as defined *ex cathedra* in 1854, regarding the Immaculate Conception of the Blessed Virgin Mary.

Through a mutual friend and bioethicist, Francis Etheredge kindly invited me to become one of his contributing authors of his excellent book titled, *The Human Person: A Bioethical Word*. This fascinating book has strongly inspired and motivated my own scholarly research into the many important beginning-of-life bioethical issues discussed in *Donum vitae* and *Dignitas personae* regarding the inviolable dignity of the *human embryo* who must be treated and cared for as a *human person* from the moment of conception until natural death.

I was surprised and deeply honored when Francis Etheredge asked me to write *An End Word: A New Beginning* for his scholarly masterpiece, *Conception: An Icon of the Beginning*. This monumental work should be given a prominent location in the libraries of every Catholic college, university, and seminary. It should be promoted and studied by everyone who hungers to know the truth about the Mystery of Christ and the Mystery of Man, created male and female by God Himself, who has eternally known and loved each one of us even before the foundation of the world. *An End Word: A New Beginning* concludes with an informative historical timeline that documents the major scientific discoveries regarding procreation - specifically *in vitro* fertilization – and includes a chronological list of key historic events in the legal battle to reverse Roe v. Wade.

The second purpose of this book – passionately shared by both authors – is to answer the crucial question: What should be done with the millions of abandoned frozen human embryos?

The final three Selected Essays in this book provide additional research and legal insights that examine several practical international and national efforts to resolve this ethical problem.

1. The first Essay was written to document the historic Alabama Supreme Court Decision that legally ruled for the first time in US history - and perhaps in world history - that frozen IVF embryos are unborn children outside of the womb, regardless of their location in a Petri dish or in cryostorage. The essay carefully documents the immediate and often unexpected national and state reactions to - and consequences of - this landmark decision.
2. The second Essay is a helpful summary of very important international and national laws that have already been enacted in other countries and here in the United States, that include policies and regulations that protect the lives of human IVF embryos from harm and death by banning experimental research on embryos, eugenic testing and selection, cryopreservation and other inhumane practices under penalty of fines and imprisonment.
3. And, finally, the third Essay, compares the development of the morality of Organ Donation with the ongoing ethical discussions regarding the morality of Embryo Adoption based upon

magisterial teachings in *Donum vitae* and *Dignitas personae* which are solidly based upon the Primacy of Human Life and the Principle of Fraternal Charity.

I would like to end this Introduction to *The Zygote of Christ and The Mystery of Man* with the well-known admonitions from Christ Himself in the Gospel of St. Matthew in which Our Lord repeatedly teaches all of us that *what we do to the least of our brethren we do to Him:*

> Then the King will say to those at his right hand, 'Come, O blessed of my Father, inherit the kingdom prepared for you from the foundation of the world. For I was hungry and you gave me food, I was thirsty and my drink, I was a stranger and you welcomed me, I was naked and you clothed me, I was sick and you visited me, I was in prison and you came to me.' [...] And the King will answer them, 'Truly I say to you, as you did it to one of the least of these my brethren you did it to me.' (Matthew 25: 40)

May we always see the faces of Jesus and Mary in every human embryo, as they are so beautifully portrayed in the *Zygote of Christ!* May we embrace Christ's words in our own hearts and ardently pray for the lives of the millions of frozen unborn children who are languishing in freezing liquid nitrogen and who need either their own parents - or adoptive parents - to come for them and lovingly welcome them into their hearts, their homes, and their forever families.

Soli Deo Gloria

Essay I

From *The Human Person: A Bioethical Word*
Pages 294 – 306

The Human Person: An Indivisibly Psychosomatic Being from Conception

Francis Etheredge

The Human Person
A Bioethical Word

ENROUTE

FRANCIS ETHEREDGE

General Introduction to Chapter 6:
The Human Person is an Indivisible Psychosomatic Whole

The modern turn to the subject has contributed afresh to the challenge to express, in a "word", what it is to be a human being: to be both capable of "subjectivity" and, at the same time, to be capable of objectifying that subjectivity. In other words, a wholesome regard for the beauty of flowers, the description of their colours and scents and indeed the thoughts which proceed from this wondrous display of developments, while common to plants generally, are an expression of the variegated subjectivity which is a natural part of each one of us. At the same time, however, as it is possible to recognize that there are common elements to our response to the activities of nature, there is a legitimate variety to it; and, therefore, objectifying our subjectivity entails recognizing that there is a certain unique unrepeatability about each response, even if, within the history of a particular person, there may be dramatic developments in the degree to which an individual responds to nature.

With the modern emphasis, too, on the identity of man, male and female, there is a further challenge to say what it is to be a human person: a human person, integrally one in soul and body, sexed, both a true individual and a person-in-relationship. On the one hand, it is true that it is possible to draw upon the ancient principles or co-principles of being, in particular, form and matter; and, indeed, it is thinking through these principles that brings us to realize both the continuity with the past and, at the same time, the need for development. On the other hand, it is perfectly valid that the times in which we live, with both its increase in understanding our embryological and neurological development, not to mention the growth of psychological and social insight, make a renewed demand on our understanding of the integrity of the human person.

Introduction to Chapter 6: Part I: An Indivisible Whole

In the investigation of human personhood it is increasingly necessary for there to be a dialogue between the traditional understanding of 'form' and 'matter' and modern embryological accounts of human conception; however, it is also necessary that the "received" theological development, particularly during the pontificate of St. John Paul II, contributes to the extent to which reason benefits from developments indicated by what exceeds its natural power. In other words, the concept of "incarnation", communicated through the theological mediation of Revelation, can contribute to a further understanding of the human person as a whole. This essay explores, then, the philosophical possibilities that arise owing to taking an "incarnational" reading of the interrelationship between 'form' and 'matter' in the context of modern embryological insights, such that traditional principles both retain their force but undergo development as they are pressed into the service of understanding human personhood as a profoundly integral whole.

In this essay, therefore, I want to consider the following: a range of starting points that open up the possibility of revisiting the ancient language of form and matter with a new possibility of their integral expression (I); various problems with the language of 'form' and 'matter' and associated terminology (II); and, finally, a new conceptualization of the interrelationship of what constitutes the human person: at once a unique personhood and being a person-in-relation (III).

A Range of Starting Points from the Work of St. John Paul II: An "Incarnational" Understanding of the Bodily Expression of the Human Person (I)

In *Familiaris Consortio,* St. John Paul II said: 'As an incarnate spirit, that is a soul which expresses itself in a body and a body informed by an immortal spirit, man is called to love in his unified totality' (11[1]). In the language he uses, which is admittedly theological language, St. John Paul II challenges us a new with how to understand the relationship, or even the interrelationship, of the human body and soul: a particular instance of the wider, metaphysical relationship between form and matter, where a particular form individuates or determines what precisely any given instance of matter is. In other words, in one sense, form and matter are "co-determining" principles of a particular object or living creature; however, these metaphysical principles express what is present in created creatures or objects as an indivisible expression of their existence. If each person, then, is 'an incarnate spirit', then it is inadequate, albeit helpful, to think of the soul as like a shadow: an inseparable "part" of the creature. An 'incarnate spirit' suggests the kind of

---

[1] The text cited is from the following location:
http://w2.vatican.va/content/john-paul-ii/en/apost_exhortations/documents/hf_jp-ii_exh_19811122_familiaris-consortio.html (published 1981).

inseparability akin to the inside and the outside of an object; indeed, there is an approximation in the expression of the concept "washing machine" in the mechanism itself. Alternatively, a more subtle expression of this principle may be a material and its properties. On the one hand, the characteristics of particles, chemical bonds, kinds of atoms and molecular combinations "inwardly" contribute to determining the texture, hardness, colour, shape and "reactivity" of materials. In other words, there is an indivisibly "functional" relationship between determining characteristics and outward properties; indeed, the outward properties express the inwardly determining characteristics of the materials.

It might be asked, then, if the co-determining principles of human being are indivisible, how is death possible? The reality of death, then, has to be understood *from the perspective of life*. Death clearly arises out of the relationship of life to the bodily expression of personhood. Thus, while ceasing to live brings about death, ceasing to live cannot rupture the existential unity of the existence of the human person. The existential unity of the person, founded by God in the very act of bringing to exist each particular person at the moment of conception, is an irreversible act. Just as it is uniquely characteristic of God to bring to exist, so the existential act of each person's existence is as irreversible as the unique power of God to bring it to exist. Whatever constitutes the mystery of death, therefore, does not constitute a dissolution of the act of existence which brings each one of us to exist; death, in its terrible reality, is nevertheless a "rupture" of the unique visible expression of the integrity of each person's existence[2].

Secondly, in *Veritatis Splendor*, he said: '*the person himself in the unity of soul and body,* in the unity of his spiritual and biological inclinations and of all the other specific characteristics necessary for the pursuit of his end' (50[3]). Again, then, we are confronted with an account of the unity-in-diversity of the human person which brings to light the integral nature of human personhood, albeit this time it is expressed in terms of the unity of man's 'spiritual and biological inclinations'. Indeed, the 'unity of his spiritual and biological inclinations' suggests the very presence of the one within the other to the point where it is not possible to say that one is not the other; in other words, while there is life, the integral unity of the human person is "indivisibly" expressed in the very structure of the interrelationship between the spiritual and the biological. Thus, just as the meaning of a word is "alive" in the context of a book but somewhat "abstracted" in a dictionary definition, so the presence of a desire for the beloved is at once integrally unitive and potentially procreative. In other words, out of the very unity of personhood arises the

---

[2] Further discussion of the nature of death is beyond the parameters of this brief article.

[3] From the text:
http://w2.vatican.va/content/john-paul-ii/en/encyclicals/documents/hf_jp-ii_enc_06081993_veritatis-splendor.html (published 1993).

ontological foundation of the inseparability of the unitive and procreative significance of the marital act (cf. *Humanae Vitae*, 12[4]).

Finally, while there are numerous other sources to draw upon, the third point of departure is from St. John Paul II's *Letter to Families,* where he says: 'When a new person is born of the conjugal union of the two, he brings with him into the world a particular image and likeness of God himself: *the genealogy of the person is inscribed in the very biology of generation*' (*Gratissimam Sane*, 9[5]). Here, what we notice, is a certain progression of insight and the view that the 'genealogy of the person is inscribed in the very biology of generation'. In other words, even in the transmission of life, St. John Paul II is articulating a personalistic understanding of the coming to be of the human person. On the one hand, there is not just inheritable characteristics embodied in the transmission of human being – but the inheritance of what parents and indeed human beings have both received and lived throughout the generations. In view of the new configuration of the inheritance received from both parents, the unique identity of the child is nevertheless an expression of the "lived" attributes of previous members of both the human family and the child's relatives. Essentially, then, inheritance is also an expression of that "translation" into actual behavioural determinants of concrete family characteristics: either of the human race or of

---

[4] From the text:
http://w2.vatican.va/content/paul-vi/en/encyclicals/documents/hf_p-vi_enc_25071968_humanae-vitae.html (published 1968).

[5] From the text: https://w2.vatican.va/content/john-paul-ii/en/letters/1994/documents/hf_jp-ii_let_02021994_families.html (published 1994).

particular expressions of our humanity. On the other hand, then, the person 'brings into the world a particular image and likeness of God himself'. In other words, it is as if the 'biology of generation' refers both to the genealogical sense of the origin of a particular person – and also to the mystery "embodied" in the very reality of the 'conjugal union of the two' bearing witness to the origin of human personhood in the very mystery of the Blessed Trinity[6]: the origin of the individual person and the community of persons (cf. *Gratissimam Sane*, 6). Thus "person" possesses, as it were, an almost impossible to describe relationality; and, therefore, while each one of us is uniquely a person, the very definition of what it is to be a person is profoundly influenced by the reality of the Blessed Trinity: the three persons in one God.

It is with this sense, then, of the integral reality of the biological, psychological and spiritual, together with the individual and the relational expression of human personhood, that we consider a certain terminological inadequacy in the terms 'form and matter' or 'soul and body'. In part, however, it may be that we are "confronting" the

---

[6] As a theological subject this is explored in various essays in the trilogy: From Truth and truth: *Volume III-Faith Is Married Reason*, published in 2016. However, there are also philosophical aspects of this question discussed in *Volume I-Faithful Reason*, published already in 2016, and in *Volume II-Faith and Reason in Dialogue*, now also published in 2016. There is also a more concerted and integrated discussion of this whole question in various chapters of *Scripture: A Unique Word*, particularly chapter twelve, drawing on embryological, philosophical and theological sources (pp. 289-336). All these works are published by Newcastle upon Tyne: Cambridge Scholars Publishing.

difficulties of a dialogue about reality: the goal of communicating not only what exists as created but how what exists as created communicates the mystery of God.

Various Problems with the Language of 'Form' and 'Matter' and Associated Terminology (II)

In the first place, there is the possibility that it has taken until the pontificate of St. John Paul II for the theological thought of the Church to mature to the point of beginning to open new possibilities to reason in the particular field of understanding the mystery of the human person (cf. *Gaudium et Spes,* 24). On the one hand, this means that there were indispensable antecedents to these developments. But, on the other hand, it means that, given the philosophical framework of perennial reflection on the reality of human personhood, upon which the work of Karol Wojtyla[7] drew, there were nevertheless "new" and innovative developments which have illuminated primordial questions on the nature of the human person. Thus the aforementioned "principles" or expressions of the

---

[7] Cf. *Love and Responsibility, Person and Act,* and the collection of essays called *Person and Community*. However, it may be said that the pontifical work (cited above) is altogether more concise, synthetic and incisively simpler than the earlier work; indeed, even to the point of it being possible that there be a development in his insight over and above the philosophical work. But it is not possible to discuss this further here.

characteristics of being, and human being in particular, are subject to a kind of theological challenge.

There are two particular problems that arise: the first concerns a latent dualism and the second concerns an insightful, but inadequate understanding of the transmission of life. Firstly, then, the co-determinants of 'form' and 'matter' had an origin in a culture in which matter was generally understood to be eternal; and, therefore, there is the implication that these "ideas" bear an implicit trace of a type of inadequate dualism[8]: the inadequacy lying in the understanding that the very co-determinants of human being were not radically "expressive" of one another[9]. In other words, matter stood in a relationship to form which made it the wholly determining principle – as if both matter and form were not subservient to a whole of which each was an indispensable "part". The analogy is too clearly that of how a sculptor shapes a stone; whereas a deeper understanding of the interrelationship of 'form and matter' is of the "whole" each contributes to express: the "idea" of the human person as

---

[8] Cf. "The Mysterious Instant of Conception", *The National Catholic Bioethical Quarterly*, Vol. 12, No. 3, Autumn 2012.

[9] This is part of the problem of an evolutionary biological development: that the body is not understood as *precisely* expressive of human emotions, intelligence and will etc. Furthermore, there is also the problem of a part being a kind of determinant of the whole. In other words, instead of understanding that a human hand is precisely what it is in view of the whole human person, it is supposed that a hand, or any other part of the human body, can lead the development of the whole: as if the whole is determined by a part – not a "part" like the "form" which determines the whole but a part that is dependent on the form for its position in the whole.

integrally one person-in-relationship. Nevertheless, let it be said, that the patristic principle that everything expresses a "divine" idea is nevertheless an intelligible expression of the view that all that exists is encompassed in the creativity of God.

Secondly, then, there is the modern understanding of the transmission of life entailing a "passage" of calcium ions which, on "contact" invigorate the ovum and by definition transform it from an inert ovum into an active human embryo[10]; and, therefore, the "transmission" of calcium ions, together with the presence and, in due course, the integral assimilation of the sperm head and its genetic content, contribute to the foundational fertilization of the ovum. The metaphysical inadequacy, then, of understanding 'form' and 'matter', *as if form wholly determines what matter to be,* in the "event of fertilization", is the very existence of the human transmission of life. An inert ovum becomes, through the very act of the transmission of life, a human embryo: the first manifestation of the presence of a human person. On the one hand, the very activity of a "flow" of calcium ions between sperm and ovum constitutes wondrous evidence of the transmission of "life-as-activity"; and, as such, opens up the whole field of the electro-chemical interactions in the animation and development of human life as a "permeating" consequence of fertilization. On the other hand, the very nature of this transmission

---

[10] Cf. *Scripture: A Unique Word,* particularly chapter twelve, drawing on embryological, philosophical and theological sources (pp. 289-336). There was also a two-part article: "A Person from the First Instant of Fertilisation? Part I", *Catholic Medical Quarterly,* August 2010, Vol. 60, No. 3, pp 12-26 and "A Person from the First Instant of Fertilisation? Part II", *Catholic Medical Quarterly,* November 2010, Vol. 60, No. 4, pp 20-26.

of life is already a "form" determined expression of matter: matter is already expressing a configured purpose in its very expression of the transmission of life. Therefore the question that arises is this: What action of God would completely accept the determinants of human life that exist and, at the same time, transform them in the "moment" of fertilization and thus bring to exist the integral conception of a human person?

A New Conceptualization
of the Integrity of Human Personhood (III)

What prompted this paper was the context[11] and thought that the theory of an emergent soul was, in fact, a plausible if unworkable account of what brings about an essentially "involved" participation of the bodily integration of the soul[12]; and, at the same time, this is another opportunity to express the almost inexpressible: the mystery of the beginning of each one of us: a mystery luminously redolent, like an icon, of the beginning.

The theory of an "emergent soul", then, is an unworkable account in so far as, in general, what does not exist cannot exist *in potentia*. In other words, the very potential of the human soul to exist cannot

---

[11] Cf. the following exchange: 2013, John Haldane, "Is the Soul the Form of the Body?", *American Catholic Philosophical Quarterly*, Vol. 87, No. 3, pp. 481-493; William Hasker, "The Dialectic of Soul and Body", pp. 459-509; John Haldane, 'Response to William Hasker's "The Dialectic of Soul and Body"', pp. 511-515; and, finally, William Hasker's, "Response to John Haldane's, "Is the Soul the Form of the Body?", pp. 517-520.

[12] Cf. William Hasker, "The Dialectic of Soul and Body".

be equivalent to the existence of the soul; indeed, it could be argued, there does exist a "type" of potentiality for the existence of the soul – but this is not to be confused with the existence of the soul in actuality. A composer possesses the potential to write a piece of music – even if the person is not exercising that talent at a particular moment; however, the potential to write a piece of music *exists precisely because* it presupposes the existence of being a composer. But the drawback to this comparison is that the existence of the person, prior to the realization of the ability to compose music constitutes, in this instance, the ground of a potential to writing music. Nevertheless, the potential to writing music still "presupposes" what exists in act, namely, the existence of a real person who possesses a real potential to write music. The kind of potential, then, that "music" possesses to being written, is all that "translates" the facility for composing into an actual composition: score; instruments; conductor and orchestra. Clearly, however, there are different kinds of contributing "potential" and they are only analogously applicable to the comparison to the potential of the "matter" to be determined by the "form" of the human soul. Nevertheless, it is useful to consider what it is about the "flesh-in-transmission" that makes it, on ensoulment, *precisely the right expression of the soul.*

An incarnational understanding of human conception

In the case of the human sperm and the human ovum, there is a potential to the "reception" of the soul; but the potential is not in view of each ovum and sperm – rather the potential to the reception of the soul is "in view", as it were, of the person who is brought to exist through the union of the sperm and ovum. Here, however, there is a further point to consider, namely this: that the immateriality of the human soul requires a concrete expression in the "materiality" of the human being. In other words, even if the human soul is immaterial, it does not follow that it does not have a material expression; and, if it has a material expression, then that material expression "stands", as it were, "*in potentia*", towards the soul's possibility of being expressed in terms of it. If human life, then, stands "*in potentia*" to the possibility of expressing the existence of the human soul, then human life, presumably, possesses intrinsic characteristics that "make it" readily receptive to the "receptive transformation" of expressing the human soul. In other words, in the very moment of the transmission of human life, when the ovum is activated by a sperm and no longer, therefore, an ovum but an embryo enclosed in an embryonic wall – there is the possibility that in that "moment" of the transmission of human life there is God's active creation and ensoulment of the human person.

On the one hand, then, there is the wholly human reality of the transmission of human life through spousal love; but, on the other hand, there is the wholly divine act of *bringing the person to exist, precisely through the transformation of the union of the spousal gametes: the union of sperm and ovum*. This "transformation" of the union of the spousal gametes is precisely what human fertilization is, as it were, such that it is truly called procreation; and, therefore, it is precisely the suitability of that union of the spousal gametes for the reception of this act of God that characterizes this moment as both truly human and truly an act of God. The plausibility, then, of the hypothesis of an "emergent soul" is this: that the first instant of human life possesses the outward characteristics capable of "expressing" the incarnation of the human soul. Furthermore, this makes apparent that the foundational moment of the existence of the human person is, in actual fact, an indivisible moment: a moment in which there was neither separable "body" nor subsequent soul; and, therefore, it is truly the case, then, that what emerges from this moment of conception is as indivisible as it is possible to be: the very existence of the human person is, *per se,* one in soul and body. The implication of this for human life is, simply, that where the body lives, there the soul is, and where both are is the person-in-relationship[13]. The very characteristics of life-in-transmission, then, in the very reality of it being a permeating "electro-chemical" activity, constitute an apt and lively expression of the outward expression of an ensouled life. In other words, just as the bodily expression of the person cannot live except that life is expressed in all the subtle forms of electro-chemical activity, so the presence of this activity and its determining of various functions is a "visible" sign of the presence of the human soul.

The transformation of life in the "moment" of the transmission of life

There are three particular characteristics which make this "moment" of human conception an outward sign of an inward act of

---

[13] A version of this principle was first elaborated in the work: Each One of Us Is an Icon of the Beginning, http://whendoestheperson-begin.info/Article8/article8.shtml

God. Firstly, the initial inertia of the ovum is radically transformed. There is now a uniquely independent, existentially existing being and activity. Secondly, the outward expression of the radical transformation of the ovum's inertia is the formation of the embryonic wall. Within the embryonic wall development proceeds according to the "logic" of human development; and, outwardly, implantation makes concrete the transitory "dependence" of the nascent human person on his mother. However, the transitory nature of human dependence is also, as it were, a permanent "image" of the interdependence of the very nature of human being and relationships, both to each other and to God. Thus, although the human person is "transitorily" dependent on his mother, we remain both profoundly dependent on God and on each other; interdependence, in other words, is a constitutive characteristic of human being. Thirdly, the sperm is the "active" agent in bringing about the transformation of the ovum's prior, inertial state; and, expressive of that change, is the "transmission of life" which occurs between what was the sperm and the ovum and is now the human embryo. Thus it is precisely as "life" that the soul animates the bodily being of the human person from conception. Life, permeating as it does the "whole being" of the bodily expression of the human person is, therefore, the perfect outward sign of the reality-presence of an animating principle: the soul or human form of the human being.

The biological "enfleshes" the psychological: the psychological is therefore "enfleshed" in the biological

Drawing, then, on the whole direction of an "incarnational" kind of thinking, the bodily expression of the person is, precisely, the outward development of the inwardly unfolding of the life of the person. What becomes clear, then, is that from the "act" of ensoulment, the existence of the ensouled body as brought about by God, there is an uninterrupted unfolding of the life of the person; indeed, this development is so intimately internal to the external manifestation of the human person, that it is possible to describe the human person's development as a biologically inscribed psychological unfolding of human personhood. In other words, the very existence of the human person, established as "act" by the ensouling action of God in the very moment of the transmission of life, is so ordered to the manifestation of the psychological reality of human personhood that, to the degree that the biological development makes possible, there is the "visibilization" of the psychological development of the human person.

Psychological development is essentially relational

Finally, just as the child comes to exist through the relationship of the spouses and to God and, therefore, relationship is fundamental to the constitution of the human person, so the biological development of the child "manifests" the essentially "relational" nature of human personhood. Psychological development, therefore, is internally developmental to the point of becoming manifest in the characteristically relational responses of the organically developing child. In other words, as soon as it is possible for the child to express himself, self-expression is psychologically inscribed in the very outward manifestation of biological development. The child's movement, then, is not just the maturation of the embryological development of visible human characteristics, it is the "visibilization" of the presence of the human person from conception. The "signs of life", therefore, so precious to expectant mothers and pensive fathers, are the concrete manifestations of the relational personhood of their child.

Conclusion

The validity of traditional philosophical principles is vindicated by their adaptability to a new understanding of them that does not, at the same time, corrupt their original meaning; indeed, true development, indicates a profound "at-one-ment" between a truly perennial philosophy and the abiding presence of the being that it explicates. In the particular case of 'form' and 'matter', the very challenge of understanding what these principles originally meant and how they arose, contributes to both a deeper appreciation of their ability to express reality-as-it-is and, at the same time, directs us to a fresh appreciation of the reality of the human person which has consistently inspired these investigations. In one sense, however, the reality of human personhood exceeds our capacity to express it; but, in another sense, there is a growing realization that recognizing that each one of us is a "person-in-relation" is taking us in the direction of a fuller understanding of the lived reality of being a human being.

Essay II

From *The Human Person: A Bioethical Word*
Pages 170 - 179

An Introduction to Genesis I-II: The Creation Gift of Participation in the Mystery of the Blessed Trinity

Elizabeth Bothamley Rex

The Human Person
A Bioethical Word

ENROUTE

FRANCIS ETHEREDGE

An Introduction to Genesis I-II:
The Creation Gift of Participation in the Mystery of the Blessed Trinity

"For You formed my inward parts;
You wove me in my mother's womb.
I will give thanks to You, for I am fearfully & wonderfully made;
Wonderful are Your works, and my soul knows it very well."
Psalm 139: 14

"It would be an honor!" I quickly responded to Francis Etheredge in early January: "I would be happy to work with you on this important book." Although we had never met - and had barely corresponded - I was already a great admirer of Francis Etheredge for his masterful and scholarly articles. "Without knowing the details of Chapter Four," I continued, "I am already intrigued by its title, and having read your article, 'Frozen and Untouchable: A Double Injustice to the Frozen Embryo' I know that we are kindred spirits in attempting to defend the very lives of the least of our brethren, especially by seeking guidance in Sacred Scripture, in the Word of God. Genesis is a gold mine!"

As I began to read and reread Francis Etheredge's Chapter Four, and then as I began to ponder and pray and plunge myself into the daunting task to provide an introduction to such an amazingly vast and profound chapter dealing with the first two chapters of Genesis, what increasingly filled my entire being - heart, mind, soul and body - was an overwhelming sense of praise and gratitude to the Blessed

Trinity. It was unlike anything I had experienced before. This time, the praise and gratitude I literally "felt" was deep and personal: it was - quite simply - overwhelmingly joyous!

As you now prepare to read Chapter Four, I urge you to pick a time and a place where you will be able to focus on every sentence – even on every word - as Francis leads you, the reader, not unlike Virgil led Dante, to "see" with your mind's eye the glorious realities contained within the first two chapters of Sacred Scripture - the Word of God – which have been divinely spoken to us by God Himself through the "literary genius" of the inspired and ancient authors of the Book of Genesis. With Francis's faithful and scholarly guidance, the attentive reader will not only explore the immense Gift of Creation itself – Heaven and Earth in all of its infinite beauty and magnificence – as well as the precious and personal Gift of Life itself, but Francis will also introduce the attentive reader to the awesome and wondrous reality of God's immense Gift of Participation – of Co-Creation – that the Blessed Trinity has entrusted to Man, male and female, who has been created in His own image and likeness.

This chapter will challenge, enrich, and expand your understanding of who "man, male and female" really "is" in the eyes of God. The Blessed Trinity, in creating Adam and Eve in His image and likeness, at the same time told them to be fruitful and multiply, inviting them to intimately participate with the Blessed Trinity, the Almighty Creator of Heaven and Earth, in God's greatest Creation of all, the conception of each and every new child – the conception of each and every new human life, the conception of each and every human being created in the image and likeness of God Himself, the conception of each and every human person; a human person who is a child of God as well as a child of Man, both male and female. Psalm 139 is so rightly full of praise: "*I will give thanks to You, for I am fearfully and wonderfully made.*"

Chapter Four is dedicated to revealing the Mystery of Man by exploring the great Mystery of God as contained in the Word of God in Genesis 1-2 that Etheredge has so beautifully "mined" for us.

Genesis is indeed an inexhaustible goldmine for the interdisciplinary field of Bioethics, and in this chapter Francis Etheredge digs deep into its first two chapters and brilliantly exposes its many bioethically-significant golden veins and nuggets for the benefit of all of us who are dedicated to unearthing, understanding, and defining the extremely elusive - and as yet undefined – Mystery of Man, male and female. Not only are we truly Children of God, created in His image and likeness, but together with God, we also participate in the very Co-Creation (the Procreation) of every Child of God.

In this Foreword, therefore, I have prepared a brief summary – a Treasure Map of sorts – as a helpful guide to search for and find some of the many brilliant passages where the reader will discover veritable treasures in Francis Etheredge's Part I of Chapter Four.

Next, I will briefly comment upon just one of Francis's many great contributions in Part II of Chapter Four, namely, how God's act of Creation communicates and makes "visible" the Mystery of God, and likewise, how the human act of Procreation communicates and makes "visible" the Mystery of Man, male and female.

Finally, I wish to offer a personal response to Francis Etheredge's very important question that he poses in Part III at the end of Chapter Four to all of his readers: "Where Do We Go from Here?"

A Useful "Treasure Map" for Part I in Chapter Four

Perhaps the best summary of the overall importance of Chapter Four can be found in the last paragraph of Francis Etheredge's General Introduction:

> The consideration that Scripture expresses that "truth which God, for the sake of our salvation, wished to see confided to the sacred Scripture" (*Dei Verbum*, 11), more than inspires a consideration of its contribution to our understanding of the human person. Perhaps one of the greatest insights arises out of pondering creation as a coherent gift of the Creator. In other words, considering creation as the work of a "craftsman" (*Wis* 13:1) leads one to think of it as a "gift-as-a-whole": a masterpiece in which everything has its proportion and place – particularly the mystery of the human person who "witnesses" to the "personhood" of the Creator and the mystery of community in the Godhead.

One of Francis's greatest insights in Chapter Four, in my humble opinion, is precisely that while the Mystery of God clearly reveals the Mystery of Man (male and female), the Mystery of Man (male and female) also clearly reveals the Mystery of the Blessed Trinity. This is perhaps *the* central theme of Chapter Four that Francis analyzes from every possible angle throughout this early and foundational chapter within his book. Francis turns first and foremost, as every faithful scholar should always do, to God's Revelation and "to a theological understanding of man, male and female" because God's Revelation, particularly the first two chapters of Genesis, again in the words of Francis, is the "place" of encounter - between God and man - "with the lasting questions of human identity."

His introductory research touches upon the key notions of "nature and change" and how "change makes a cause 'visible'" because "change is a 'witness' to an agent of change" as the movement of leaves in a tree is a "'witness' to the presence of wind or other force." Thus, he concludes, "the very existence of creation is a 'witness' to a radical act of God: 'In the beginning, God created the heavens and the earth." (Gn 1:1)

The next stop on the "treasure map" in Part I is a thorough and detailed discussion of the relationship between the Word of God and human authorship.

In Part I, I found particularly intriguing Francis' commentary and comparison of the unobserved beginning of the universe with the frequently observed beginning of human life. Francis was very interested in the fact that while no one ever witnessed the beginning of the universe, nevertheless he realized that it had already occurred to the ancient author(s) of Genesis that the beginning of a person was an act of God. He pointed out that the inspired author(s) explained that after Adam and Eve came together as husband and wife and conceived a child, Eve was quoted as saying, "I have gotten a man with the help of the Lord" (Gn 4:1). Francis clearly, then, made the important point that even back then "in the intuition of the author, and perhaps already in the cultural understanding of the beginning of human life, there is a recognition that human personhood requires an act of God."

Francis realized that the author of Genesis had already clearly understood that if God can "bring a man to exist, it is possible that He can bring everything to exist." In other words, "if there was a time when a 'person was not', perhaps this helped to found the faith … that 'nothing is impossible to God'" (Lk 1:37) and that "God was capable of the event of creation" – literally out of nothing - "before there was even nothing in existence." While such a concept is so familiar to us today, it took experience, wisdom, and faith to reach such an amazing conclusion so long ago. Thus, Francis provides us with an excellent example of how the Mystery of Man can and does reveal the Mystery of God and His creation of heaven and earth.

Another treasure follows with a lively discussion of "three possible responses to the opening chapters of Genesis: unreal; unacceptable; or literary genius." Francis dismantles the first two theories and forcefully argues that "the early chapters of Genesis are a brilliant summary of salvation history, expressing the clear understanding that everything points to a beginning;" a beginning that communicates "the power, order and love of God in the act of creation."

The final treasures in Part I involve an important discussion regarding Creation as "idea" or as "reality" or as "idea embodied in reality." Francis makes the point that the "Creator of the universe *brings about the inextricable union between idea and reality*" and that there is "an indivisible relationship between idea and object." Why must we know this? Because, as Francis warns us, it is possible to "falsify what we are and therein enters the lie," and we can also "deliberately 'miscommunicate' what exists" which, in turn, can lead to several errors including fundamentalism.

The section titled "Faith or fundamentalism" is of great importance to the proper understanding of Genesis. This section clarifies that it is faith that "illuminates the complex interplay between the historical events and the authorship which is both rooted in history and transcends it." Francis also brilliantly proposes that it is faith that "enables us to persevere with the difficult questions that the Scripture poses and, at the same time, to recognize that the word of God is always relevant to each person's life and to the times in which we live."

Ultimately, Francis explains, "the Love that God expresses in everything that He does unfolds that original Love in word and deed: a deed illuminated by His word and a word fulfilled in His works."

Part II of Chapter Four – Human Procreation: Man's Co-Creation with the Blessed Trinity

The second part of Chapter Four provides an invaluable journey through and a renewed appreciation of the "marvellous prose-poetry of the opening words and chapters of Genesis." And, for those of us who have never studied Hebrew, Francis unveils the incredible depth and richness – and significant gender-related overtones – of the original Hebrew words that English translations simply cannot ever adequately convey.

As a non-Hebrew literate reader myself, I was simply amazed and delighted as Francis carefully and meticulously explains in this section the "truly divine-human" – the "living language" – of the opening Hebrew words of Genesis: *"In the beginning God created the heavens and the earth."* (Gn 1:1)

Take it slow, read and reread every paragraph. Guided by Francis' extraordinary exegesis, let the magnificence and significance of the first words of Genesis lead you to explore the personally plural depths of God Himself and of Genesis's "two different accounts of creation" which reveal God's act of creation as a "continuous whole," as "an action begun but not finished," and as "a work both complete and ongoing" in which man – male and female – is invited to participate in its dominion and entrusted to share in its stewardship by God Himself.

At this point, Francis set out into the deep to explore and "mine" the wondrous creation of man, male and female, which he refers to as a very personal and even Trinitarian relationship with God: *"God created man in his own image, in the image of God he created him; male and female he created them"* (Gn. 2:21-22).

Francis first dwells on the three different Hebrew names for God that are used in Genesis which he views as a "significant indication of the real presence of three distinct but related uses of the divine Being: God; Spirit of God; and thirdly, the Lord God" (cf Gn 2:4). Francis then proceeds to compare this implication of "the interrelationship of one divine being to another" with the "dynamically structured account" of the creation of Adam "who is fashioned by the Lord God from the ground 'and' the breath of life" followed by the "procession of Eve 'through and from' Adam."

Secondly, and majestically, Francis proceeds to discuss the extraordinary "sign-value of human sexuality" – of man being created male and female – as the first early figure of salvation history as a whole: as God's own "marriage of the Lamb" (Rev 19:17) between "Christ the Bridegroom and His Church the Bride" (cf Rev 19: 6-9).

Finally, at the end of Part II, in the section titled "The dialogue of the sexes: dominion, identity and difference in God and man," Francis dwells specifically on Genesis 1:26, where God, speaking in the plural, says "Let us make man in our image, after our likeness; and let them…." This passage in Genesis compares God's eternal "unity-in-diversity" with the "unity-in-diversity" relationship that God created "between the man and the woman" who were created in His image and likeness.

Francis Etheredge's fascinating and awe-inspiring journey through the first two chapters of Genesis culminates with a final and important discussion regarding the transcendent and personal difference between man, male and female, and the rest of creation. "Adam," he points out, "is specifically differentiated from the animals" due to "the very intimate involvement of the Lord God in the forming of Adam and Eve" by the breathing of "the breath of life" into Adam's nostrils. As written in Genesis 2:7, "man is a 'living being' *precisely because of the animating breath of God."*

In summary, as brilliantly discussed by Francis, the literary genius of Genesis portrays the first man, Adam, as lovingly created by God from the dust of the earth *and* the living breath of God; and the first woman, Eve, as lovingly created by God from flesh that was taken "close to Adam's heart." But that was only "the beginning" of the ongoing mystery of man, male and female: for God "blessed them and… said to them, 'Be fruitful and multiply, and fill the earth …'" (Gn 1: 28). Never again did God directly create a man, either male or female, without "the mysterious communion" between the man and the woman who were blessed and told to be fruitful and multiply "through the reciprocal gift of self which God has brought about through the created dialogue of man, male and female."

The procreation of each one of us, and of every human child that has ever been conceived since the creation of Adam and Eve, is an action that "is more intimate to the beginning of life than we can understand." It was Eve herself, in Genesis 4:1, who first thanked and praised God following the birth of their first son, saying: "I have gotten a man with the help of the Lord." The first woman was profoundly aware of the "very presence of God" deep within the mystery of human procreation: Adam and Eve knew they were "fearfully and wonderfully" made and blessed to participate and co-create with God.

Thus, as this chapter is so aptly titled, human procreation is "The Creation-Gift of Participation in the Mystery of the Blessed Trinity."

Part III of Chapter Four – Where Do We Go From Here?

The third and final part of Chapter Four is a call to each one of us, the readers, to ponder – deeply – the opening words and chapters of Genesis that proclaim God's infinite love for man, male and female, and for each and every child of God who is conceived as the fruit of their marital embrace.

Together with Francis, who offers his own answer to his important question, we are called to meditate upon the Word of God as Moses and God's people have done forever, throughout salvation history. Why? Because the mystery of God reveals the mystery of man, and likewise, the mystery of man reveals the mystery of God.

It has truly been an honor to read and comment upon Chapter Four. I have tried to practice what I have preached, namely to ponder the many and great truths that are contained in the opening words and chapters of the Word of God in the Book of Genesis.

In answer to Francis's important question at the end of Chapter Four, I believe we must unite and create a worldwide effort to defend and "choose life" for the "least" of the least of our brethren – the smallest of us all – the fully human, amazing, one-cell human zygote.

Is a tiny, one-cell human zygote really a human person though, with a body and an immortal soul created in the image and likeness of God? "We don't know," some readers may be tempted to answer.

If you are unsure, then let me ask this question: was Mary a person with a body and soul at her Immaculate Conception? "Of course she was," one will surely answer. And was Jesus, the Second Person of the Blessed Trinity, was He really Perfect God and Perfect Man at the moment of His Incarnation, even though He was just a tiny, one-cell, human zygote? "Of course," one will answer again, "It must be true!"

Therefore, these two divinely revealed truths regarding the Immaculate Conception of Mary and the Incarnation of the Son of God reveal these important truths regarding the absolutely amazing moment of human conception, when we are conceived as tiny zygotes.

"And the King will answer them, "Truly I say to you, as you did it to one of the least of my brethren, you did it to me." (Mt 25: 40)

I wish to sincerely thank Francis Etheredge for his gracious patience with me in submitting this Foreword beyond the desired deadline. Perhaps it was providential, however, since it was finally finished today, Saturday, May 13th, 2017, on the 100th Anniversary of Our Lady of Fatima.

May the Blessed Virgin Mary intercede for us, she who is praised in today's *Magnificat* readings as "the most perfect, the greatest, the holiest human being ever to live," – the very "pinnacle of God's creation"[14] – and who was "full of grace" from the moment of her Immaculate Conception as a tiny, one-cell, wondrous human zygote.

The Canticle of Mary

"My soul proclaims the greatness of the Lord,
my spirit rejoices in God my Savior,
for He has looked with favor on his lowly servant.

From this day all generations will call me blessed:
The Almighty has done great things for me,
And holy is his Name" (Lk 1:46).

---

[14] Rev. Peter John Cameron, O.P. (Editor), "Saturday, May 13, Our Lady of Fatima," *Magnificat* 19, no. 3, (2017), 170.

Essay III

From *Conception: An Icon of the Beginning*
Pages 508 – 535

The Gift and Rights of Being Conceived In-Relationship: We Are an Icon of the Beginning

Francis Etheredge

CONCEPTION
AN ICON OF THE BEGINNING

FRANCIS ETHEREDGE

✦ENROUTE

The Gift and Rights of Being Conceived In-Relationship: We Are an Icon of the Beginning

The Help of Faith and Reason: Christian Dogmas, Conception and the Gift of Completing a Homeless Child's Embryological Development

We live between where we were and where we are going to be; and, in a sense, that defines our whole reality: a person is what I was at conception, who I am and what I am becoming. In the biblical sense, I am an unfinished work between being conceived and being resurrected from the dead; and, therefore, the whole nature of being a person encompasses both everything to do with beginning and everything expressed in the mystery of rising, God willing, with Christ. A biblical conception of person, therefore, takes account of a trans-temporal understanding of the human person and, in a word, takes us into the realm of the spiritual transformation of all that I am and all that we are: a radical transfiguration of human being.

At the same time, however, we live and work out our salvation in the present; and, in that respect, the being we are to become is also expressed in all its on-going trans-temporality. There are many ways of perceiving the human person and it is possible to view ourselves "fractionally", according to the lens of one discipline or another; however, it is necessary to strive for an adequate account of the whole human being, even amidst the kaleidoscoping fragmentation which is so often what becomes of our self-perception. This fragmentation shows itself most tragically in the reductionism which renders one person the "object" of another person's intrusive investigation, manipulation or destruction; and, therefore, there is an urgency in seeking to recover an account of the "whole" that we are which includes our relationship to God, to others and to each other.

Thus it is time to seek and to advance an integral account of the beginning of our whole human nature. At the same time this offers reasons, both religious and realistic, to recover the perception of the humanity of the frozen human embryo and to advance the possibility of housing the "homeless child" in the hospitality of a mother's nurturing womb and the relationships which stem from that beautiful fact. As a whole, while it is both urgent and timely to act for the good of another, it also necessary to widen the debate and to call for a new global ethic which defines the integral individual and social requirements of human development from the first instant of conception onwards. Thus, in this final part of Chapter Five, the resources of both reason and Revelation are brought to bear on the exciting, challenging and ever urgent question of the beginning of the human person and the way forward for the most neglected of our brothers and sisters.

Introduction to Chapter 5: Part III: Beginning an Unfolding of Human Personhood. The complementarity of faith and reason derives, ultimately, from their origin in the being of the Blessed Trinity and the "act of gift" which brought the universe and, in particular, the human race into existence. This is not a discussion, however, of the vast subject of our original beginning and development;

rather, it is a specific exploration of the "moment" of human conception (cf. Gn 4: 1) and, in the light of more recent challenges, of the help a "homeless" human embryo needs.

As we shall see, natural and divine truths are both complementary and ordered to one another. While, however, there remains a certain difficulty in proving that conception is the first intimately instant moment that a human being begins to exist there is, nevertheless, the brilliantly illuminating embryological evidence between mother and child collected by Professors Justo Aznar and Julio Tudela and, as we have also seen, the assistance of Scripture, Tradition and the Magisterium of the Catholic Church. On the one hand there are those who hold the view of delayed ensoulment: that God gives the gift of a human soul at a point subsequent to the first instant of fertilization. On the other hand there are those who see that the evidence "itself" is advocating an increasingly convincing position of a first instant of human conception: a "moment" in which God gives the integral gift of human personhood; indeed, the moment in which the divine-human giving of life is expressed in the whole gift of human personhood: of a being-in-relation – a child. In other words, although there is, as it were, body and soul and the possibility of a "two-fold" giving of human life, it is argued here that the very nature of human wholeness entails an inseparable act of human existence in which God gives there to be "one" being in the first instant of coming to exist as a bodily-expressed-soul. Thus there is the first task of establishing a coherent account of the beginning of each one of us. This author will then advance the position of an "immediately" enfleshed and animating human ensoulment of the human being; and, in addition, that this originating beginning of a particular human person is simultaneously the commencement of all social and spiritual relationships.

On the basis of this brief exposition, it will then be considered what light it is possible that the divine mysteries of the *Immaculate Conception, Incarnation and Eucharist*, can contribute to understanding how to help the "homeless" human embryo. This is not primarily, then, a discussion of the immorality of *in vitro* fertilization or any other procedure which brings about the conception of a human being outside of the marital embrace; rather, the focus of this part of the discussion is on the very existence of a "homeless" human embryo requiring the completing nurture of mothering and the dynamic of human parenting. Thus this author will advance the position that the very deficit of a mothering embrace, which ordinarily begins when a man's wife becomes pregnant, already implies a "need" to rectify the injustice to the child of being "homelessly" conceived. In other words, it is not that this child does not have the parents who contributed to the existence of the child; but, in the very nature of being "instrumentally" conceived, the child is "relationally impoverished" in terms of the very dynamic characteristic of human development: the participation of the "whole" of human parenthood. The utterly poignant situation of the completely incomplete reality of being conceived without the ensuing completion of natural human development is epitomized in the triple tragedy of the freezing of an embryonic human child: of being conceived outside the natural complementarity of the womb; of being conceived in a way that obscures the reality of the "person-as-gift"; of "freezing", as it were, the inherently progressive developmental goal of manifesting the whole humanity of the child. On what basis do those who enjoy the full benefit of human development deprive others of it?

There are a number of parts, then, to this article; however, there are the following five main divisions: What is Conception? (I); Conception in the Teaching of the Church (II); Conception in the Mysteries of Mary, Jesus and Each of Us (III); the Morality of Embryonic Transfer and Adoption (IV); and the Adoption of the Embryonic Human Child (V)[15].

What is Conception? (I)

Conception, ordinarily, is a beginning. The beginning of a beginning is, as it were, the first instant of that beginning. Therefore the conception of a human being is from the first instant of fertilization. In view, however, of the natural "uncertainty" that "surrounds" the possibility of human conception, it is clear that there is a complementary psychological disposition to the act of being open to the gift of human life: of an integrally grateful giving which both belongs to the marital embrace and "opens" upon the possibility of a child. In other words, even in an account of the "moment" of conception it is necessary to recognize that this belongs to the intimacy of marriage and the language of love's giving; indeed, it is possible that a reason for the very "objectification" of the conception of the

---

[15] I acknowledge a particular debt to Dr. Elizabeth Rex for her encouraging and stimulating feedback in response to a prior draft of this part of the work; and, therefore, I am grateful to what has led to the re-crafting or further development of a number of earlier points.

human person is the tendency to separate what belongs together: to divorce human conception from the marital act. The popular expression, "making love", bears the intuitive insight that belongs to a wealth of human experience: that love entails the good of the lover and the loved. The integral fruit of the reciprocal good of the love that becomes husband and wife is the good, nascent but begotten in the love of each other, that unfolds and develops in the "passing" from the possibility to the actuality of conceiving a child.

Furthermore, almost as a kind of natural theological outcome of philosophical speculation on the beginning of human personhood, there is a threefold argument for the action of God at the first instant of conception. What is beyond the nature of the contributing factors must come from an agent more capable that they are: therefore a soul is created and ensouled by God. Secondly, the very nature of an instant beginning of an everlasting human person implies a uniquely inner determination of what constitutes the ripe moment of the whole beginning of personhood. Thus an ensouling instant that begets a beginning is expressive of the power of God. Thirdly, God so perfects human cooperation in that inward moment of ensoulment being manifest in an outward beginning of embryonic human life that it evidences His completely intimate "involvement" in the mystery of human conception: a truly human-divine act of begetting a child of man, male and female, and a child of God.

The assumption that the human embryo is not ready to be "ensouled" at the first instant of fertilization presupposes an ancient account of the human soul as actually bringing about changes in the "matter": as if the changes in the matter were "dependent" on a threefold process of ensoulment. Thus the plant-like level of life required a vegetative soul. Vegetative development matured and was then followed by the requirement of an animal-like soul stimulating sensitive development. Following the maturing of an animal-like development the process culminated in a readiness for the reception of a rational soul and the wholeness of human being[16]. In other words, implicit in this understanding is the view that rational ensoulment requires a specific developmental stage of "bodily" being. However, the human body is not defined by the absence of the human soul but by its presence, except at death; and, therefore, delayed ensoulment points to the need for a more integral understanding of the "whole" of human being.

The reality of the first instant of the sperm's animation of the ovum is expressed by the enclosure of the sperm head "in" the embryonic wall. In other words, up until the penetration of the ovum by the sperm, there are two entities that are nevertheless ordered to each other in the transmission of human life; but, on contact, the inert ovum is animated by the active sperm and becomes, in that instant, a human embryo. The inertia of the ovum is evidenced in the inactivity of the mitochondria: the energy centres of the ovum-as-cell. The formation of the embryonic wall, then, constitutes an outward sign of the new reality of the child conceived[17].

---

[16] Cf. Chapter Twelve: Life "from" Life: A Reflection on the Moment a Person Comes to Exist, pp. 289-336, particularly page 304 of Scripture: A Unique Word, Newcastle upon Tyne: Cambridge Scholars Publishing, 2014).

[17] This summary statement is based on prior research, which can be found in Chapter Twelve: Life "from" Life: A Reflection on the Moment a Person Comes to Exist, pp. 289-336, particularly pages 317-322 of

In answer to the main objection to delayed animation, modern embryology makes it very clear that the whole embryo is marvellously coordinated from the beginning, developmentally independent of both parents, while naturally dependent on the mother for psychologically embracing embryonic nurture and, thirdly, the child is an integrated whole whose goal is the progressive manifestation of being the person he or she is. Thus there is no obstacle to ensoulment from conception, by God, which is also a kind of "incarnation" in that the soul constitutes the person, one in body and soul (cf. *Familiaris Consortio*, 11); indeed, if the soul is understood as the "life" of the person, not only does the soul constitute the integrating principle of personhood but it is, too, the new "life" of the person who comes to exist through the union of the gift of parenthood. Not only is the soul the integrating principle of human life, evidenced by the disintegration which follows death – the human soul is also inseparably one with the matter which constitutes the human being: man is one in soul and body (*Gaudium et Spes*, 14). In other words there is an indescribable intimacy to the union of soul and matter which brings the human person to exist just as surely as the musician's breath brings forth a note from a musical instrument. Just as the saxophone is "dormant" without the breath of a musician, so the first instant of fertilization needs the personalizing presence of the human soul. In other words, it is abundantly clear that the "matter" involved in the transmission of human life is uniquely derived from the man and the woman and is, therefore, intrinsically adapted to

---

Scripture: A Unique Word). But see, too, the middle part of Chapter 5 of this book, by Professors Justo Aznar and Julio Tudela.

the reception and expression of the animating human soul. The human inheritance is, therefore, a deeply rooted reality which is totally taken up in the transmission of human life and the procreation of a child. There is not, then, the "imposition" of a "form" on matter; rather, there is an inward change which manifests the personally humanized inheritability of the human race. On the one hand, then, there is an act of ensouling animation in the moment that the bodily expression of the person comes to exist; and, on the other hand, there is an ensouling animation of what constitutes the bodily expression of the human soul. The "matter" of the body is not, therefore, some kind of "indifferent" substance; rather, in the words of St. John Paul II, the matter of the body is 'the genealogy of the person ... inscribed in the very biology of generation' (*Letter to Families*, 9).

Conception in the Teaching of the Church (II)

We need to begin by recognizing that although the Church has not committed herself to an affirmation of a philosophical nature concerning the moment of ensoulment there is, nevertheless, a tendency in her documents to refer to conception as the beginning of the presence of the human person. A careful consideration, however, of the wording of these documents, particularly *Donum vitae, The Gift of Life*, raises a twofold possibility.

The first possibility is that conception is assimilated to the unification of what were the separate nuclei of sperm and ovum. In the English translation of *Donum vitae* there are two points to note: the first is a reference to a beginning and the second defines that beginning to be when *the nuclei* of sperm and egg have fused. Thus we read in the main text of *Donum vitae*: 'the fruit of human generation, from the first moment of its existence, that is to say from the moment the zygote has formed, demands the unconditional respect that is morally due to the human being in his [or her] bodily and spiritual totality' (*Donum Vitae*, I, 1). In the English text, then, there is the following defining note which goes on to say: 'The zygote is the cell produced when the nuclei of the two gametes have fused'[18]. Thus *Donum vitae* could mean that the first instant of personal individuality is the first instant of the unification of the nuclei of the sperm and the egg[19].

The Latin text, however, does not include the defining note which says, 'The zygote is the cell produced when the nuclei of the two gametes have fused'; rather, the Latin text of *Donum vitae* simply says that the zygote comes to exist *"orta a fusion"* (arising from a fusion)[20]. Thus the Latin text could mean that the zygote comes into existence from the first instant of the fusion of the sperm and the egg which, as we know, developmentally unfolds uninterruptedly from then on; but, taking the sense of 'arising from a

---

[18] *Donum vitae*,
http://www.vatican.va/roman_curia/congregations/cfaith/documents/rc_con_cfaith_doc_19870222_respect-for-human-life_en.html.

[19] This whole discussion of the different translations of *Donum vitae* is more extensively carried out in Francis Etheredge, *The Human Person: A Bioethical Word*, St. Louis, MO 63109: En Route Books and Media, 2017, pp. 361-389: Chapter Five: On the Development of the Church's Prudential Judgements on Human Conception and the Plight of the Frozen Embryo", particularly page 369.

[20] Etheredge, *The Human Person: A Bioethical Word*, p. 369.

fusion', it could be argued that the English was simply making more explicit what was in fact the sense of the Latin, namely, that 'fusion' is more of a "step" in the process rather than just a first instant. Therefore, it could be taken that the English note explicates the Latin phrase, "*orta a fusione*' (arising from a fusion)'. In other words, in the English translation of *Donum vitae* conception of the person is assimilated to the first instant of the fusion of the nuclei; but, in the case of the Latin, which is the traditionally more authoritative text, the meaning is more open and could include the very first instant of fertilization.

Thus the second possible meaning of conception is the prior, first instant of fertilization referred to earlier, namely, the sperm's animating penetration of the ovum and the formation of the embryonic wall. In other words, the "delayed animation" type of understanding of conception is that of a "moment" subsequent to the embryo's animation by the sperm's penetration of the ovum; and, while possibly coherent with Church teaching, has the obvious drawback of presupposing an almost dualistic "combination" of "soul" and "body". The original meaning of conception, however, is that of referring to a real beginning of an actual entity; and, therefore, there is the possibility that the second meaning of conception, more coherent with its actual meaning, is where the truth is leading us. Thus the integrity of a being formed, whole and entire from the first instant of conception, is both more coherent, consonant with the facts and, it is argued, "falls" within the range of meaning expressed in the Latin text of *Donum vitae*. In other words conception 'arising from a fusion' can denote, specifically, the ovum's reception of the sperm or the sperm's penetration of the ovum which, together and in the "one" moment, "effect" the formation of the embryonic wall.

Conception in the Mysteries of Mary and Jesus and Each one of us (III)

In the following two sections it is necessary to establish what can be known, both naturally and supernaturally, concerning the beginning of each one of us.

The fact of human conception (IIIi)

What can be reasonably established about human conception is that there is an obvious start to the new entity of the human embryo: the formation of the embryonic wall on the penetration of the ovum by the sperm. Delayed animation of the embryo is more "interpretative" of the facts than the more self-evident reality of there being a radically new beginning for a new human being; indeed, it could almost be argued, for relationships to be real there has to be a point at which a child radically exists as "present" to the parents. What better moment, then, than the first instant of "love's" manifestation of the coming together of husband and wife!?

The reciprocal relationship between natural and revealed truth (IIIii)

On the basis of a real, outward sign of the origin of human personhood in the "sperm-inclusive-enclosing" of the embryonic wall, there is a natural "sign" capable of expressing an "inner mystery": the natural outward sign of the enclosing embryonic wall expressing the inner moment in which God determines there to be an animating human soul. Thus we can argue that God acts in a way which gives witness to His action, not because He needs it but because it is a part of the Creator's communication to us of the nature of human being and the mystery of God. Thus creation is a witness to the act of the Creator who brought it about; and, similarly, pouring water over the head of an infant, together with the Trinitarian words of the minister, is an outward sign of the gift of baptism. Just, then, as a sacrament is an "outward sign" of an inward action of God, so the formation of the "embryonic wall" is an outward sign of the inward action of God that brings a person to exist[21]. In other words, God reveals Himself through 'deeds and words' (*Dei Verbum*, 2[22]) which, taken together, are like a hermeneutical principle: a principle through which to understand God as Creator, Redeemer and Sanctifier. Therefore, even in the instance of a natural sacrament, like the first instant of fertilization, God acts in a visible way to communicate the invisible reality of bringing the person to exist – the whole person, one in body and soul (*Gaudium et spes*, 14).

Secondly, there is a kind of "Patristic Principle" based on a number of texts of the Fathers of the Church which, in effect, lead to the conclusion that whatever is true of human conception is redeemed by the coming of Christ; and, therefore, as the nature of human conception becomes better understood, so it is clearer that redemption

---

[21] Cf. Etheredge, *Scripture: A Unique Word*, Chapter 12.

[22] I am aware that the text of *Dei Verbum* does not apply, specifically, to the beginning of human life; however, in so far as it applies to the action of God generally, it applies specifically to the action of God at conception.

begins with the beginning of the Conception of Mary and the *Incarnation* of Christ[23]. While, then, God generally acts in a way that communicates His mystery to the creatures He has created, this particularly applies in the conception of Mary. God's conception of Mary is an ensouling action of God which takes up the wholly natural contribution of her parents into the history of salvation. Thus the conception of Mary was, in effect, an act of God as Creator, Redeemer and Sanctifier; and, therefore, it could be said that although human conception is a "deed and word" which would not normally be directly implicated as an act of saving love, on reflection we see that in fact as human conception entails an act of God it is almost intrinsically ordered to salvation history.

There are acts of God in the Old and New Testaments which express, more directly, conception as ordered to salvation history, particularly the conceptions of Isaac, Samuel and John the Baptist. More widely, then, expressing the truth that each person comes to exist in relationship to Christ is indeed to realize that the action of God at conception is in the heart of salvation history; as it says in *Gaudium et spes*: 'For, by his incarnation, he, the son of God, has in a certain way united himself with each man' (22). If, then, it is true of human conception in general that this is an integral expression of how we are "begotten" in salvation history, then how much "more" true is this of the mystery of the *Incarnation* in which, uniquely, the Son of God is expressed in the flesh of human being. Just as the 'Son of man' (Mt 20: 28) is hypostatically united to God in Jesus Christ, and each one of us is 'in a certain way united' with Christ, so each

---

[23] Cf. Etheredge, *Scripture: A Unique Word*, Chapter 12.

one of us is, as it were, "in potentia" to the possibility of participating in that hypostatic union. Perhaps, in one sense, baptism and the other sacraments more generally, are an "actuation" of this mystery of our participation in the hypostatic union between God and man in Jesus Christ.

Thirdly, what can be recognized as true of human conception will apply to Mary's *Immaculate Conception*, which was to be humanly conceived in a state of original grace; and, moreover, what follows from her graced conception assists us to understand the first instant of her conception and, therefore, the first instant of our conception. In the mystery of Mary's *Immaculate Conception*, the radical gift of God's grace brings forth the integrity of the woman who is to bear the Christ-child: the mystery of Mary's graced-nature informs our understanding of the conception of each of us. In other words, if Mary is to be free of the "taint" of original sin, the original deprivation of graced-nature which Adam and Eve both lost themselves and passed that "loss" on to us, then it follows that from the first instant that her flesh existed she was conceived without original sin.

Moreover, as it was put by Blessed John Henry Newman, grace is 'a real inward condition or superadded quality of the soul'[24].

---

[24] The quotation is cited from *Difficulties Felt by Anglicans*, Vol. II, p. 46, published on pp. 22-23 of *Mary: The Virgin Mary in the Life and Writings of John Henry Newman*, edited with an Introduction and Notes by Philip Boyce, Leominster, Herefordshire: Gracewing Publishing, 2001.

Given, then, that a personal grace requires a personal subject[25], it would follow that Mary was "present" from the first instant of her conception; and, in view of what we know of human conception, the first instant of human conception is the formation of the embryonic wall following the sperm's penetration of the ovum. The dogma of the *Immaculate Conception*, it could be argued, while advancing an extraordinary grace for Mary in terms of being conceived without original sin builds on nature to do so[26]; and, in building on nature to do so, the dogma of the *Immaculate Conception* is an implicit confirmation of there being a first instant of human conception from

---

[25] References to the following work were suggested by the Rev. Dr. Richard Conrad, OP, (email, 10/3/2019): Cf. St. Thomas Aquinas, *Summa Theologiae*, Methuen, 1992: I-II, Qu 113, Art 10: 'the soul has a *natural capacity for grace being made in the God's image*' (p. 321); I-II, Qu 63, Art 2: If the standard of virtue 'is God's law then … [it] can only be caused by an activity of God within us' …. [furthermore, taking account of the possibility of the presence of original sin, it is possible to say with St. Thomas that 'Such divinely instilled virtue cannot co-exist with mortal or fatal sin' (p. 241); I-II, Qu 5, Art 5: 'a nature that can thus achieve utmost perfection, even though needing external help to do it, is of a nobler constitution than a nature that can only achieve some lesser good, even though without external help' (p. 181); I-II, Qu 1, Art 8: 'Men attain their goal by coming to know God and love him' (p. 174); I, Qu 93, Art 4: 'grace adds to some men an actual if imperfect understanding and love of God' (p. 144) which, in the case of the Blessed Virgin Mary, goes to the limit of human perfection and excludes even the possibility of original and personal sin; I, Qu 8, p. Art 3: 'God exists in those actually knowing and loving him, or disposed to do so; and since this is God's gracious gift to reasoning creatures we call it existing by *grace* in his chosen friends' (p. 22).

[26] "Grace builds on nature" (St. Thomas Aquinas).

which her redemption follows[27]. While Mary's redemption is different from ours in that it is from the first instant of human conception, Mary's human nature is common to ours in that it is from the first instant of human conception[28]. Mary, then, after Adam and Eve, is the pre-eminent case of the human person united to the Son of God; and, in so being, it makes radical sense that she is wholly without sin and completely human, one in soul and body, from the first instant of her conception. In other words, the redemptive mystery that the Son of God, 'has in a certain way united himself with each man', takes on a particularly transparent "completeness" in view of Mary being sinless from conception; and, on that basis, radically unites her to the salvific work of her Son, Jesus Christ, the Son of God made man. Just, however, as the Blessed Trinity is at work in the creation of man, male and female, so the Blessed Trinity is present in the re-creation of the human race through the *Incarnation* of Christ: 'The angel announced to her not just the incarnation but fundamentally the entire mystery of the Blessed Trinity ...'[29].

Fourthly, in the case of the *Incarnation* of the Son of God there is the "originality" of the gift of the second Person of the Blessed Trinity being enfleshed in the womb of Mary. While the eternal Son of God becoming man is a truly extraordinary event, the very

---

[27] There is a more complete account of the argument summarized here in Francis Etheredge, *A Little Book on Mary and Bioethics*, a work in progress (forthcoming from enroutebooksandmedia).

[28] Cf. Etheredge, *Scripture: A Unique Word*, Chapter 12.

[29] Cf. Hans Urs von Balthasar, Mary for Today, Middlegreen: St. Paul Publications, 1987, translated from the original German, p. 35.

"conception" of Christ entails a first moment of the fertilization of the human ovum as it is animated by Christ's human soul and, therefore, begins an unfolding development which expresses the personhood of the Christ-child. In other words, just as the conception of the Virgin Mary entailed a first instant from which "she" was conceived, one in body and soul, so there is a first instant from which Christ was miraculously conceived, one in body and soul; and, if a first instant, then the very first instant of human conception if Christ's soul was not to be united to flesh "tainted" with the "loss" of original sin. On the one hand, then, the Son of God is God from God and eternally "of" the Father; but, on the other hand, being "of" Mary in the very nature of human temporality, it follows that He was conceived in a way which united Him to the very depths of human conception: depths from which our redemption proceeded as from the deepest origin of each one of us. If then Christ is God from God uninterruptedly, then it follows that Christ is man from Mary in the first instant of "becoming flesh" (Jn 1: 14).

Fifthly, our understanding of the Eucharistic "presence" of Christ is of an instantaneous change of bread and wine into the *Body and Blood of Jesus Christ*; and, therefore, there is the principle of an "instantaneous" change existing, manifesting itself and being an expression of the action of God. An action of God at conception, then, is no less complete and instantaneous than in the "moment" of changing the bread and wine into the *Body and Blood of Jesus Christ*. In other words, just as the universe did not always exist but was given existence, so a human person did not exist and is given existence. Thus the gift of each human person is in the context of God giving, from all eternity, the gift of creation existing from the moment of His action; and, therefore, there is a kind of "echoing" between " the beginning" (Gn 1: 1) and the "conception" of each one of us: an "Iconic" reverberation between the one and the other: an act of God at conception that "makes present" the beginning of everything in the new beginning of each person – all of which speaks of the "immanent presence of God". Just as having spoken is an irrevocable act, so is the beginning of existence an irrevocable change which conditions the development of what unfolds from it: A human being constitutes a relational whole which is both immersed in the whole history of the human race and, at the same time, is a new beginning of the matrix of relationships in which each one of us is both immersed and expressed.

It is clear, then, from the aforementioned natural and supernatural arguments that there is a "congruence" between natural and supernatural truth in determining the first instant of human conception. The implication of this conclusion is that there is a vital necessity, then, to an action which seeks to save a child conceived – if conception is from the first instant of fertilization. Is there, then, a similar congruence between faith and reason in determining the help necessary to the "homeless" human embryo? "Home-less", not because of being "un-housed", but "homeless" because of being conceived without the nurturing home of the mother's womb.

The Morality of Embryonic Transfer (IV)

There are two sections to this part of the article: the first section concerns a brief exposition of the natural arguments for Embryonic

Transfer and Adoption (IVi) and the second section concerns the supernatural arguments for Embryonic Adoption (IVii).

Natural Arguments for Embryonic Transfer and Adoption (IVi)

At its simplest, a child conceived outside of the womb is without the immediate possibility of benefitting from the mother's nurturing contribution to the completion of embryological development; and, therefore, there is a natural injustice to the child who, in "his" own way, cries out for redress: a voiceless cry which appeals to us the more it is almost inaudible in being submerged in freezing temperatures. This is the unnatural drama of the ordinarily "invisible" reality that has been brought into existence by others and is an ongoing "rupture" in the universality of the human right to the completing nurture necessary to human beings at the beginning of life; indeed, this human right belongs, inexorably, to the human right to the gift of life once given: the gift of life is an irrevocable gift and entails the whole manifestation of the person conceived. In other words, the child is first and foremost a gift: a gift each one of us is given to be; and, in the case of a frozen human embryo or other abuse of the recipient of this gift, there is an injustice done to the very being of the child. At the same time, there is obviously a problem in perceiving the humanity of the child in storage; and, indeed, perhaps the sophistication of the methods of preserving the child in storage is an indication of the lengths to which "we" will go to "hide" the humanity of the frozen child.

Secondly, then, having been conceived in a way contrary to the very humanity of natural relationships, this child is dependent on being given hospitality in the womb of a woman; and, in order to make this possible, the child must be transferred to her womb from a glass dish or a place of storage. The transference of the frozen child to the womb of the mother confronts the reality of the event of freezing a child: that this is another human being, equally gifted with the gift of existence as you and I.

This procedure of transferring the embryo is radically different from the artificial methods of conception and transportation used up to this point. The reason that the transfer of a child to the womb of a woman is not to be confused or assimilated to *in vitro* fertilization and its methods is that this child now exists; and, intrinsic to his or her existence, is the natural right to completing human development which, in the case of the embryonic stage of a child's development, requires the nurturing presence of a woman willing to be an adopting mother. The natural object of the act of embryonic transfer, then, is that of taking the embryonic child and placing "her" in the womb of a woman; and, as an integral part of this process, following through on the adoption procedure which would help the maturing identity of the child. Embryonic transfer is therefore helping to rectify the relational deficit that was incurred in the very nature of the mechanical method that was employed to conceive the child.

The adopting woman and, by implication, her husband and their family are giving the humanitarian aid that this child needs to live; and, in this essential respect, they are fulfilling the natural gift of womanhood, parental care and family experience that is indispensable to this child's growth and maturation[30].

The Supernatural Arguments for Embryonic Adoption (IVii)

Just as in the case of understanding human conception, the question arises as to the help we may derive from considering the nature of the Christian mysteries particularly, in this case, the mystery of the *Incarnation* of the Son of God; clearly, however, there is an important and irrevocable distinction: the Christian mysteries are acts of God for our salvation, whereas conceiving a child outside the spousal embrace is already abandoning the implicit requirements of relational conception and development. The possible help to us of the mystery of the *Incarnation* of the Son of God is recognizing that a humanitarian act that is different to the natural order is not necessarily contrary to it. The Church has already indicated that embryo transfer is acceptable when it is for the good of the embryo and, therefore, the possibility that embryo adoption is also acceptable for the same reason: it is necessary for the good of the homeless embryonic child[31].

It has already been stated that there is an injustice to the child conceived without the natural possibility of completing his or her

---

[30] Cf. Etheredge, various parts in *The Human Person: A Bioethical Word*, particularly Chapter 7, Parts IV-V.

[31] Cf. Dr. Elizabeth Rex, "IVF, Embryo Transfer and Embryo Adoption", NCBQ, Summer 2014; and cf. Etheredge, *The Human Person: A Bioethical Word*, Chapter 7: Parts IV-V.

course of mothering nurture. God, however, even in view of this injustice, has given the gift of personal life; and, therefore, it could be argued, God is not responsible for the injustice to the child conceived but is, as it were, responsible for the life He has given. God, in His salvific acts, is constantly communicating salvation to the human race; and, in one respect, God's saving acts are always in the context of man's prior sin. Thus man's prior sin is not an obstacle to God's saving acts but, rather, the "occasion" of God showing a love greater than the death of sin. God acts for the life and salvation of each one of us, even amidst the tragedies of sin and disorder which arise out of immoral and unnatural human action. In the particular case, then, of the frozen embryonic child, a child whom God 'has in a certain way united [to] himself'[32], it is argued that the very adopting love of the rescuing husband and wife are a concrete expression of redeeming human love: a redeeming human love that is what it is because of the action of God within it that reaches to the needs of the child that was frozen.

God commands us to love the least of our brethren; and, in this situation of an orphaned embryonic child, the requirement of love is expressed in meeting the needs of a radically homeless child: a child conceived with a "deficit" to the right to the completing natural development of the manifestation of his or her personhood. The focus of God's salvific act, then, as always, is not the evil or injustice committed by human beings but what is necessary to remedy the harm that they entail. Thus, in this instance, it could be argued that the natural law expresses the will of God in that we do what we are

---

[32] Slightly adapting *Gaudium et Spes*, 22.

able to help the orphaned child. At the same time, however, the adopting husband and wife express the gratuitous nature of redeeming love: that just as there is a gratuitous act of giving life – so there is a gratuitous act of saving life. Steering free of all commercial and quality control entanglements, therefore, is essential to the adopting husband and wife's involvement in the rescue of this child; indeed, doing so, purifies the relationship between adopting parents and child and "returns" the child to the gratuitous love from which being conceived "artificially" artificially removed him or her.

The *Incarnation* of the Son of God is an act according to the nature of God the Father's eternal generation of the Son and, through the action of the Holy Spirit, is also according to the relational nature of human conception; it is certainly an extraordinary act in terms of it occurring within the marriage of Mary and Joseph without entailing their marital embrace and indeed implying the preservation of their virginal love of God and each other. In the case, then, of the adoption of the embryonic homeless child, aided as it is by transferring the child to the woman's womb, it is an act of life-giving charity which delivers an innocent child from the possibility of the indeterminate frustration of normal human development, disablement or death by deterioration; and, while this act of adoption occurs, of its nature, outside the context of an original act of spousal love, in providing for a homeless embryonic orphan this adoption begins to communicate the very nature of "redeeming" love as the gift to the child of the good necessary to his or her life.

In a word, just as God's gift of life is completely gratuitous, so an act of redemptive love is completely gratuitous; and, just as God gives human life according to the covenant of the flesh He founded, so His saving acts are according to the needs of the human life conceived. Nevertheless, even the gratuitous nature of redeeming love acts in accordance with the natural law that expresses our human participation in the divine law; and, therefore, what is done to rescue an illicitly conceived child is completely different to the action which caused the child's embryological, developmental and relational "homelessness": the injustice expressed in the conception of a "maternally homeless" child is addressed by the justice of an indispensably generously gratuitous adopting love. In a word, just as redemption goes beyond original sin without endorsing it, so an adopting love goes beyond the injustice of a child conceived "maternally homeless" without endorsing the method through which the injustice was perpetrated.

The Adoption of the Embryonic Human Child (V)

Conception, then, begins a biologically inscribed psychological development that unfolds inseparably socially and spiritually. Just as God acts at the beginning of each one of us, so His action begets the beginning of the spiritual relationship which unfolds in terms of the whole of life and is lived, intensely, in prayer and communion with others. From the beginning, the whole "bio-physiological psychological dialogue" between mother and child is, as it were, in the lived context of spousal love and human fatherhood. The "presence" of the personal communication between mother and child[33] is, then,

---

[33] Cf. Foreword and Biography to Chapter Five by Kathleen Sweeney, *The Human Person: A Bioethical Word*, pp. 226-233.

in the presence of the husband and father – even if this is at times not as possible as it is for the mother. The very personal nature of "mothering" starts to "make visible" that there are personal relationships at the root of human being: the personal relationships which express an interpenetrating parental and divine love which seeks the flourishing of "who" has begun to exist as the fruit of love.

In a sense, then, one of the most challenging tasks of our time is to recover the reality of human conception as an expression of love; and, at the same time, to recover the reality of the person conceived as gift from gift: the gift of the child from the reciprocal gift of human love: the reciprocal gift of human-divine love. It seems all too possible to traverse the myriad paths to an ancient truth and yet never to arrive at it so vividly as Eve did: "I have gotten a man with the help of the Lord" (Gn 4: 1). It is true that there are many wonderful developments in the world around us and in the culture that permeates our everyday world. In one sense, however, there is a failure of reason and imagination - an inability to recognize that each one of us is a living witness to an astounding fact: that the frail temporality of our beginning belies an immutability of "who" comes into existence. God, in His unbounded generosity, "recognizes" a person in and through an "ensouling" act whatever constitutes the real beginning of a human life: that whatever the imperfection of the circumstances of human conception, if it is the conception of a human being, then he or she is a person begotten and beloved by God unto the possibility of eternal life.

Perhaps what we need is the contemplative complement[34] to the analytical approach: to pause in front of our own identity, the wonder of our children and indeed the mystery of the Christ-child; indeed, in terms of the modern roots of our thinking it may be necessary to revisit the question of the *whole* of human being[35]: an understanding of the mystery of human being as proceeding, as it were, as a whole from the contemplative gaze of God from all eternity. It may even be that we are facing a failure of faith, too, in grasping that if God gazed from all eternity on man, 'male and female' (Gn 1: 27) contemplating, as it were, the creation of human being in the light of His own mystery, that He saw a perfectly whole human personhood: completely integrated from the first instant of fertilization, relational and wondrously manifesting the nature of personal being! Just as we need to enliven our perception of each existing person and the first "moment" of his or her existence so we need to fall again in prayer in front of the mystery of our Creator, Redeemer and Sanctifier.

---

[34] On the back cover of *The Human Person: A Bioethical Word,* Rev. Dr. Nicanor Austriaco says that Etheredge's set of essays 'emerge from a contemplative reflection upon what it means to say that the person is a 'created word'".

[35] This emphasis on the "whole" of human being arose, in a way, from considering various insights in the work of Edith Stein, "A Gift from Edith Stein (1891-1942): http://www.hprweb.com/2017/09/a-gift-from-edith-stein-1891-1942/; but now a part of the published book, *The Family on Pilgrimage: God Leads Through Dead Ends*: http://enroutebooksandmedia.com/familyonpilgrimage/.

In conclusion, there is a congruence of reason and faith in both understanding the original moment of human conception and what, therefore, constitutes the good action which expresses the "redeeming" love of an orphaned embryonic child. In a word, the adoption of a radically "homeless" embryonic child, conceived from the first instant of fertilization, is an act of redeeming love: an act that participates in the mystery of God's adopting love of the human race (cf. Eph 1: 5).

Progress in the Church's teaching, then, not only proceeds in terms of the relatively recent discoveries concerning the nature of human conception; but, in addition, can be further enabled by the "organic" dialogue between the dogmas of the Christian Faith and their "implied" foundation in the mysterious fact of human existence. Thus it is necessary to argue for a more coherent exposition of the dogma of Mary's *Immaculate Conception* and the nature of every human conception: the action of God from the first instant of fertilization.

Just as creation and redemption require a radically original moment of human conception, so do human and divinely redeeming adoption reciprocally illuminate each other as gratuitous gifts of love; and, if adoption is the work of the redeemer, and grace builds on nature, then the radical adoption of a "homeless" embryonic person is yet a further expression of the lengths to which love goes in the "homing" of the homeless. Thus this essay argues for the clarification of the teaching of the Church on the right to embryo adoption: the reciprocal right of offering and being offered the possibility of embryo adoption[36].

In a word, then, if there is progress in understanding the irrevocable moment of human conception then it beholds the human community to articulate this truth in a newly formulated charter of human rights; indeed, from what begins as a whole comes the unfolding of what began from the first instant of that beginning. As truth and goodness are inherently ordered to one another, then the whole gift of human personhood entails the immutable right of the full unfolding of what has begun. 'If we are all equal in the receipt of the gift of human life'[37] then the unfolding of that gift is a common good. Thus the human person, the human being-in-relation, the child, has the right to an integrally human conception, development and manifestation of the human person. Thus this book argues for a new instrument of human rights to be debated and clarified for the benefit of the whole human race.

---

[36] Cf. Conclusions: 'Regardless of the above comments, we [are] of the opinion that the statements in Dignitas Personae offer no settled moral assessment of embryo adoption. Thus, we certainly believe that there is no impediment to continue investigating the moral foundations of this practice, until such times as the Catholic Church issues a definitive moral judgement on it' (Acta Bioethica 2017; 23 (1): 137-149, p. 147 of "Moral Assessment of Frozen Human Embryo Adoption in the Light of the Magisterium of the Catholic Church", Justo Aznar, Miriam Martinez-Peris and Pedro Navarro Illana: http://www.bioethicsobservatory.org/wp-content/uploads/2018/07/Moral-assessment-of-frozen-human-embryo-adoption.pdf.

[37] Etheredge, *The Human Person: A Bioethical Word*, p. 347.

The Reality of two Children: Spencer and Caroline

"Spencer and Caroline were conceived during *in vitro* fertilization (IVF) and were cryopreserved for some years before being lovingly placed for adoption through the Nightlight Christian Adoption Agency and its Snowflakes Embryo Adoption Program. Over 650 "Snowflake Babies" have been adopted as frozen human embryos since 1998. For more information please visit: www.snowflakes.org. (Permission was granted to use this photograph.)"

Essay IV

From *Conception: An Icon of the Beginning*
Pages 585 – 643

An End Word: A New Beginning

Elizabeth Bothamley Rex

CONCEPTION
AN ICON OF THE BEGINNING

FRANCIS ETHEREDGE

⊕ENROUTE

An End Word: A New Beginning

"The human being must be respected as a person – from the very first instant of his conception."~ Donum vitae

Introduction

I have chosen this beloved portrait of Saint John Paul II to introduce this End Word because it is only fitting that this magnificent work should also *end* as it *begins* on its first page:

> *"This book is a thanksgiving to God for the spiritual fatherhood of the late Pope St. John Paul II; and more generally, for the ways that papal writing and the ministry, the work, and the presence of many others have enriched my life."*

Conception: An Icon of the Beginning by Francis Etheredge is a profound, scholarly, and masterful presentation of the most important scientific, philosophical, theological, scriptural, and magisterial evidence that explores, defines, and answers the single most important question facing the entire human race today: When does human life begin?

It was a great – but very humbling – honor to be asked by Francis Etheredge to provide an End Word. When I pleaded with him for some guidance, his advice made perfect sense:

> "Do not worry about an extensive review of the book. The point is to look ahead and around, not back; although historically, in terms of the **reality of human life**, we do need to **look back in order to look ahead.**
>
> But as regards the book, **only refer to what is in the book so that you can go forward.**"
>
> Go, therefore, for what is manageable and, as you say, **indicate new developments, whether in the law or elsewhere.** A bit like the end of a PhD thesis, it is important to **acknowledge where work needs to be done** and, if you see clearly, **to end with some bold points about these possibilities**"[38].

Therefore, inspired by Francis Etheredge's comments, this End Word will first *"look back in order to look ahead"* by providing an important historical timeline that begins with the relatively simple and fairly unnoticed developments that led from the scientific

---

[38] Email on 7/18/2018, with emphasis added.

discovery of fertilization to the current myriad of highly significant scientific, ethical, legal, and magisterial events *"in terms of the reality of human life."*

For many of us who are not professional research scientists, it may come as a surprise to learn that the scientific discovery of fertilization did not occur until 1843. Regardless, however, of our field of scholarly expertise, we must learn from history and take seriously the prohibition to experiment on any human being, regardless of his or her age or stage of biological development. Medical and technological advances must be therapeutic and ethical at all times. Why? Because, as it is so carefully and thoroughly discussed in this outstanding book, every human conception is an icon of the beginning; every human being is a child of God whose spiritual soul is directly and immediately created in the image and likeness of God at the transformative first moment of human conception.

Finally, in order to *"acknowledge where work needs to be done,"* I will briefly refer to just five of the many significant contributions made by Francis Etheredge in order to "end with some bold points about these possibilities."

A LOOK "BACK"

A History of Human Fertilization & Conception From 1818 to the Present

We must "look back in order to look ahead...
in terms of the reality of human life."
~ Francis Etheredge

Louise Joy Brown is the world's first test-tube baby, and her birth in 1978 was considered to be a major milestone in the history of what is now called ARTs, the acronym that is used for Assisted Reproductive Techniques. Her birth was heralded as the beginning of a new era that would be filled with promising new technological advances in the service of human life.

But as history has so often demonstrated, technology itself, if not restrained by ethical regulations and by the law, can also be used as a powerful weapon to dehumanize, manipulate, and destroy human life. Sadly, Aldous Huxley's warnings, in his 1932 futuristic novel Brave New World, have already become a global reality that is in desperate need of ethical boundaries and strict international regulations before it is too late.

Louise Joy Brown was born in England on July 25, 1978, after being conceived in a Petri dish on November 10, 1977, during a procedure which is now known as *in vitro* fertilization or IVF. Today, Louise lives in the UK with her husband and their two sons who were born in 2006 and 2013.

By 1990, just twelve years after her birth, roughly 90,000 IVF babies had been born. In 2008, a worldwide study prepared by the International Committee for Monitoring Assisted Reproductive Technologies (ICMART) estimated that over 5 million babies had been born using IVF; and by 2018, just 40 years after the birth of Louise Brown, ICMART estimated that 8 million babies had been born using IVF. What is rarely reported, however, is that tens of millions of other human embryos perish during IVF because they were often discarded, donated to destructive embryonic research, or remain frozen in cryostorage tanks filled with liquid nitrogen.

Who, when, where, why and how did all this happen? While it would be impossible to list every relevant event and development, the following historical timeline nevertheless presents a history of some of the major developments in every discipline regarding human conception and fertilization. I am very grateful to the following web site and its timeline for documenting many, but not all, of these important events: http://www.pbs.org/wgbh/ americanexperience/babies/timeline /index.html/

1818 - 1978: From *Frankenstein*, to the Scientific Discovery of Fertilization, to the First Test-Tube Baby

1818 – On January 1, the first edition of the novel Frankenstein; or, The Modern Prometheus was published anonymously. It was written by 18-year-old Mary Shelley and tells the story of a young scientist, Victor Frankenstein, who creates a monster as a scientific experiment in his laboratory.

1827 – Karl Ernst von Baer discovers that the female body contains egg cells called ova (plural) or ovum (a single egg).

1833 – On July 26, the Abolition of Slavery Bill passed in the House of Commons in England, 3 days before the death of William Wilberforce, who had led the long parliamentary campaign to end slavery throughout the British Empire.

1843 – Martin Berry, a physician, discovers that fertilization occurs when a single sperm penetrates the ovum.

1854 – On December 8, the Apostolic Constitution *Ineffabilis Deus* is promulgated *ex cathedra* by Pope Blessed Pius IX. The Dogma of the Immaculate Conception of Mary settles a 600-year debate regarding the first and second instance of Mary's Conception. "[Our Predecessors, the Roman Pontiffs] never thought that greater leniency should be extended toward *those who, attempting to **disprove** the doctrine of the Immaculate Conception of the Virgin, **devised a distinction between the first and second instance of conception*** and inferred that the Conception which the Church celebrates was not that of the *first* instance of conception but the *second. In fact, they held it was their duty* not only to uphold and defend with all their power the Feast of the Conception of the Blessed Virgin but also *to assert that **the true object of this veneration was her conception considered in its first instant (in primo instanti)***"[39].

1855 – Woman's Hospital opens in New York City. Over the next two years, its chief doctor, Dr. J. Marion Sims performs artificial insemination 55 times during which he injected sperm from the husband into the uterus of the wife. Only one pregnancy was ever achieved, and it ended in a miscarriage.

---

[39] Emphasis and italics added. Cf. http://www.newadvent.org/library/docs_pi09id.htm.

1856 – On March 6, the U.S. Supreme Court infamously ruled (7-2) in Dred Scott (slave) v. Sanford (owner) that Scott was not a citizen based upon racial arguments that were enforced by those "who held the power." Regrettably, it was Chief Justice Roger B. Taney, a devout Roman Catholic, who delivered the majority opinion: "The question before us is whether the class of persons described in the plea in abatement compose a portion of this people, and are constituent members of this sovereignty? We think they are not, and that they are not included, and were not intended to be included, under the word "citizens" in the Constitution, and can therefore claim none of the rights and privileges which that instrument provides for and secures to citizens of the United States. *On the contrary, they were at that time considered as a subordinate and inferior class of beings who had been subjugated by the dominant race, and, whether emancipated or not, yet remained subject to their authority, and had no rights or privileges but such as those who held the power and the Government might choose to grant them….*"[40].

1863 – On November 19, President Abraham Lincoln delivers his famous Gettysburg Address, in the midst of the Civil War

---

[40] Emphasis added to the citation from: https://billofrightsinstitute.org/dred-scottv-sanford-1857-excerpts-majority-dissenting-opinions.

that ultimately ends slavery. Lincoln reaffirms the Nation's dedication *"to the proposition that all men are created equal."*

1865 – On January 31, the 13th Amendment to the United States Constitution abolishes slavery in the United States.

1868 – On July 9, the 14th Amendment to the United States Constitution is ratified. It legally establishes that "States may not deprive any person of life, liberty, or property, without due process of law; nor deny to any person within its jurisdiction the equal protection of the laws."

1876 – Oskar Hertwig observes the fusion of spermatozoa with ova (of a starfish) for the first time.

1884 – A doctor in Philadelphia, Dr. William Pancoast, performs the first known case of artificial insemination by donor (not by husband), by injecting sperm from a medical student while the woman was under anaesthesia. Nine months later a baby boy was born, but Dr. Pancoast never told either the husband or the wife what he had done.

1909 – 25 years later, an article about artificial insemination by donor appeared in the *Medical World Journal*, and Dr. Pancoast was severely criticized.

1926 – On January 30, the American Eugenics Society (AES) was formally incorporated to promote eugenics education in the United States with the goal of improving the genetic composition of humans through controlled reproduction. Irving Fisher from Yale was its first president. In 1972, AES was renamed the Society for the Study of Social Biology.

1926 – Funded primarily by the Rockefeller Foundation, the Committee for Research on Problems of Sex is founded. For the next 20 years, this well-funded Committee does research on reproductive endocrinology, i.e. reproductive hormones.

1928 – The ovarian hormone *progesterone* is discovered by scientists. The role of progesterone is to thicken and enrich the lining of the uterus each month in order to receive and nourish the embryo. Following the embryo's implantation in the womb, progesterone continues to be produced in the placenta during the pregnancy.

1929 – The ovarian hormone *estrogen* is also discovered. Estrogen affects the development of the female body and prepares her genital tract for fertilization, implantation, and the nutrition of the human embryo. Too much or too little estrogen can adversely affect the female body. Birth control pills contain estrogen and affect ovulation and implantation.

1932 – Novelist Aldous Huxley publishes the futuristic novel *Brave New World* that portrays a society that is primarily comprised of test-tube babies. This shocking novel continues to affect the ongoing debate regarding assisted reproduction.

1944 – Dr. John Rock at Harvard and his lab assistant Miriam Menken conduct successful human *in vitro* fertilization experiments but they do not implant any of the embryos. Their IVF research generates public interest and concern.

1947 – On August 20, the International Medical Tribunal in Nuremberg, Germany, convicted 23 Nazi doctors for crimes against humanity. The tribunal issued a set of 10 international rules for all future human experimentation. The first rule of the Nuremberg Code states: **"The voluntary consent of the human subject is absolutely essential"**[41].

1948 – On December 10, the Universal Declaration of Human Rights (UDHR) was proclaimed by the United Nations General Assembly in Paris as a common standard of achievements for all peoples and all nations. It sets out, for the first time, fundamental human rights to be universally protected and is considered a milestone in human history. Article 3 states, "Everyone has the right to life, liberty and security of person"[42].

---

[41] Emphasis added to the citation from: https://history.nih.gov/research/downloads/nuremberg.pdf.

[42] https://www.un.org/en/universal-declaration-human-rights/.

1949 – Pope Pius XII publicly condemned the fertilization of any human ova outside the body of a woman. Those who do so, he said, are taking "the Lord's work into their own hands."

1961 – An Italian scientist, Daniele Petrucci, fertilized 40 human ova. He allowed one embryo to develop for 29 days and even attain a heartbeat before destroying it. The Vatican publicly denounced his experiment as "sacrilegious."

1968 – An American doctor, Robert Edwards, and an English gynaecologist, Patrick Steptoe, agree to work together using a new abdominal surgery technique, called laparoscopy, to retrieve a mature human egg for *in vitro* fertilization.

1968 – Pope St. Paul VI issues a papal encyclical called *Humanae vitae* that forbids the use of artificial contraception because it immorally violates the "inseparability principle" between intercourse and procreation. While IVF is not mentioned, the "inseparability principle" equally applies to IVF because it also separates intercourse and procreation.

1969 – A Harris poll claims that a majority of Americans believe that IVF is "against God's will."

1970 – New York State legalized abortion and stunned the Nation[43].

---

[43] https://www.nytimes.com/2000/04/09/nyregion/70-abortion-law-new-york-said-yes-stunning-the-nation.html.

1971 – At a Washington, DC conference on medical ethics, the Noble laureate James Watson, who collaborated in the discovery of the double helix structure of DNA, warns the conferees that IVF research necessarily involves infanticide. Dr. Edwards (mentioned above in 1968) stood up, publicly defended his IVF research, and was given a standing ovation.

1972 – The American Medical Association (AMA) urged a *moratorium* on all IVF research involving humans; but the American Fertility Society urged further work.

1972 – On April 16, ten thousand New Yorkers gathered in Central Park to protest New York's 1970 abortion law. After the protest, the NYS legislature votes to repeal the law; but one man, Gov. Rockefeller (R), vetoes the will of the people.

1973 – On January 22, the U.S. Supreme Court legalizes abortion in Roe v. Wade. The majority decision uses 13th century theology and science to support their erroneous decision about when human life begins, stating: *"Christian theology and canon law came to fix the point of animation at 40 days for a male and 80 days for a female,* a view that persisted until the 19th century…. *Due to continued uncertainty about the precise time when animation occurred, to the lack of any empirical basis for the 40-80 day view, and perhaps to* **Aquinas'** *definition of movement*"[44]. Anti-abortion advocates also opposed IVF research and experimentation that involved the deliberate destruction of human embryos.

1974 – A couple in New York City sues Columbia-Presbyterian Hospital for $1.5 million dollars when the hospital deliberately destroyed the couple's IVF embryos. The hospital feared it would lose its government grants.

1976 – English gynaecologist, Dr. Patrick Steptoe, meets with Lesley and John Brown. Lesley has blocked fallopian tubes. The doctor recommends *in vitro* fertilization to them.

1977 – On November 10, Dr. Steptoe surgically removes one egg from Lesley's ovary and fertilizes it in a petri dish. After two days, the eight-cell IVF embryo is transferred into Lesley's uterus. In December, her pregnancy is confirmed.

1978 – Anticipating a media frenzy, the Browns sell the rights to their story to a British tabloid for half a million dollars.

1978 – On July 25, the birth of Louise Brown makes headline news around the world and raises legal and ethical questions.

---

[44] Roe v. Wade, 410 U.S. 113 (1973) IV.3, with emphasis and italics added.

1978 – On October 16, Cardinal Karol Josef Woytyla is elected as Pope John Paul II. He is inaugurated on October 22.

1978 – 1998: 20 Years After the Birth of the First IVF Embryo, the First "Adopted" Frozen Embryo is Born

1979 – Following discussions by bioethicists and theologians regarding the ethics of IVF and the moral status of human embryos, the Ethics Advisory Board of the then-Department of Health, Education and Welfare of the United States published a document recommending a 14-day limit to the growth of a human embryo *in vitro*. It became known as the "14-Day Rule" and several countries pledged to prohibit *in vitro* experimentation on human embryos beyond 14 days, i.e., the 14-day old embryos involved would all be destroyed.

1986 – On August 29, Lewis E. Lehrman, a distinguished businessman, philanthropist, writer, historian, and Lincoln scholar publishes, *The Right to Life and the Restoration of the American Republic.* His compelling article explains that America "was founded 'under God,' begotten as Thomas Jefferson wrote, according to the 'Laws of Nature and of Nature's God, a nation dedicated, in fact, to a religious proposition, a principle of natural theology." As such, the "unalienable right to life is not, for America, a single issue, but a first principle, a self-evident truth established at its Founding.... The truth is that life, liberty, and the pursuit of happiness are a logically ordered sequence. The rights to liberty and to the pursuit of happiness derive from every man's right to his own life and are meaningless without it"[45].

1987 – On February 22, *Donum vitae* is promulgated during the pontificate of Pope St. John Paul II. It forbids IVF and all other Assisted Reproductive Techniques. But it also declares as "*licit*" and "*desirable*" therapeutic procedures that are carried out on the human embryo that "are directed toward its healing, the improvement of its condition of health *or its individual survival*"[46].

1990 – Germany, in compliance with the Nuremberg Code, passes the world's first "Embryo Protection Act." Under penalty of fines and/or imprisonment, human embryos may not be used for scientific experimentation, harmed, or killed[47].

1990 – The Human Genome Project (HGP) begins to research is the DNA sequence of the entire human genome.

1993 – On August 6, as a response to pervasive and global moral relativism, Pope John Paul II promulgates *Veritatis*

---

[45] https://www.crisismagazine.com/1986/the-right-to-life-and-the-restoration-of-the-american-republic.

[46] *Donum vitae* I, 3. Emphasis added.

[47] https://www.rki.de/SharedDocs/Gesetzestexte/Embryonenschutzgesetz_englisch.pdf?__blob=publicationFile.

Splendor, a profound encyclical that reflects upon the solid moral foundations of the teachings of the Catholic Church.

1995 – On March 25, Pope John Paul II promulgates the encyclical *Evangelium vitae: on the Value and Inviolability of Human Life*. It defends the sanctity of life and prohibits abortion at any stage of human development *from the first moment of conception*. It prohibits contraception, *in vitro* fertilization, and all destructive research on human embryos.

1996 – A female sheep called "Dolly is the first mammal to be cloned but the controversial news is not revealed until 1997.

1996 – The U.S. Congress bans federal funding for research on embryos through "the Dickey-Wicker Amend-ment that prohibits the use of federal funds for the creation of human embryos for research purposes or for research in which human embryos are destroyed, discarded or knowingly subjected to risk of injury or death greater than that allowed for research on fetuses in utero"[48].

1997 – On August 15, the *Catechism of the Catholic Church* is approved and promulgated during the pontificate of Pope St. John Paul II. It states, "Since it must be treated from conception as a person, the embryo must be defended in its

---

[48] https://www.researchamerica.org/advocacy-action/issuesresearch-america-advocates/stem-cell-research/timeline-majorevents-stem-cell.

integrity, cared for, and healed, as far as possible, *like any other human being*"[49].

1998 – Hannah Strege, the world's first adopted frozen embryo is born in San Diego, California on December 31, 1998. Nightlight Christian Adoptions, a fully licensed adoption agency located in Santa Ana, California, assisted Hannah's adoptive parents, Marlene and John Strege, with the adoption of several frozen embryos who were relinquished for adoption by a loving placing family.

1998 - Present: Scientific Evidence Reveals the Beginning of Human Life at the First Instant of Fertilization

1998 – On November 1, the Human Fertilisation and Embryology Act 1990 is approved by the Parliament of the United Kingdom. Its official title states that it is "An Act to make provision in connection with human embryos and any subsequent development of such embryos; to prohibit certain practices in connection with embryos and gametes; to establish a Human Fertilisation and Embryology Authority; to make provision about the persons who in certain circumstances are to be treated in law as the parents of a child; and to amend the Surrogacy Arrangements Act 1985." Among

---

[49] CCC, n. 2274, with emphasis added. CCC, n. 2275, also cites *Donum vitae* I, 3 (Cfr. 1987 above).

its many policies, it limits the storage of human embryos to a maximum of five years.

1998 – On November 9, the UK passes the Human Rights Act. It compels all public organizations, including the Government, police and local councils, to treat everyone equally, with fairness, dignity, and respect. Article 1 defends the Right to Equality, and Article 2 defends the Right to Life.

2003 – In January "Dolly," the cloned sheep is euthanized due to premature arthritis, lung disease and premature aging.

2003 – In April, the Human Genome Project is completed.

2003 – A Harris poll finds that a majority of American believe that infertility treatments should be covered by insurance.

2004 – Italy passes a Medically Assisted Reproduction Law, also known as Legge 40, that strictly regulates the fertility industry. A maximum of three IVF embryos may be created and every embryo must be transferred. Pregnancy reduction is prohibited. The cryopreservation of human embryos and destructive embryonic experimentation are both prohibited. Frozen embryo adoption is legally permitted and encouraged[50].

2005 – On April 2, Pope St. John Paul II dies in Rome.

---

[50] https://www.ieb-eib.org/nl/pdf/loi-pma-italie-english.pdf.

2005 – With the Vatican's support, Italy's Medically Assisted Reproduction Law survives a public referendum to repeal it[51].

2006 – The National Catholic Bioethics Center and the Westchester Institute for Ethics and the Human Person jointly publish *"Human Embryo Adoption: Biotechnology, Marriage and the Right to Life."* Edited by Rev. Thomas V. Berg, L.C. and Dr. Edward J. Furton; and Foreword by Robert P. George.

2006 – On July 19, President George W. Bush vetoes the Stem Cell Research Enhancement Act. Many *"Snowflake"* children, i.e., children who were adopted as frozen embryos, and their adoptive parents attend the White House ceremony.

2007 – On January 16, a previously frozen embryo named Noah is born in New Orleans. Following Hurricane Katrina, a cryostorage tank containing 1,400 frozen embryos was rescued from a flooded hospital on September 11, 2005. Noah's rescue and birth 16 months later make national news.

2007 – Dr. Maureen Condic, a researcher and professor of Neurobiology and Anatomy, publishes the article, *Life: Defining the Beginning by the End* in the scholarly journal, *First Things*. Dr. Condic scientifically argues that "Embryos are not merely collections of human cells…. Embryos are genetically unique human organisms, fully possessing the

---

[51] https://www.nytimes.com/2005/05/31/world/europe/in-political-step-pope-confronts-law-on-fertility.

integrated biologic function that defines life at all stages of development, continuing throughout adulthood until death"[52].

2007 – Editors Sarah-Vaughan Brakman and Darlene Fozard Weaver publish, *The Ethics of Embryo Adoption and the Catholic Tradition: Moral Arguments, Economic Reality, and Social Analysis.* The publisher is Springer.

2008 – Authors Robert P. George and Christopher Tollefsen publish, "*Embryo: A Defense of Human Life.*" They argue that no one should be excluded from moral and legal protections on the basis of age, size, or stage of biological development.

2008 – During the pontificate of Pope Benedict XVI, the "Instruction *Dignitas personae* on Certain Bioethical Questions" is officially dated September 8, 2008, the Feast of the Nativity of the Blessed Virgin Mary, but it is not publicly released by the Congregation of the Doctrine of the Faith until December 12th at a Vatican press conference. *DP* begins by teaching that "[t]he dignity of a person must be recognized in every human being from conception until natural death" and that the "teaching of Donum vitae remains completely

---

[52] http://www.firstthings.com/article/2007/01/life-defining-thebeginning-by-the-end.

valid, both with regard to the principles on which it is based and the moral evaluations which it expresses"[53].

2008 - At the Vatican press conference, the President of the Pontifical Academy for Life tells the *Catholic News Service* that "the discussion is still open" on the matter of embryo adoption and specifically *states that "the Vatican did not rule out the practice"*[54].

2008 – On December 9, the United States Conference of Catholic Bishops releases a summary that states that while *Dignitas personae* "raises cautions and problems" about proposals for the adoption of abandoned frozen embryos, it *"does not formally make a judgment against them"*[55].

2012 - On December 31, The Telegraph reports that more than 3.5 million human embryos have been created between 1991 and 2012: approximately **1.4 million embryos** were transferred to the womb, with only about one in six resulting in a pregnancy (16%); about **1.7 million** embryos were discarded; about 270,000 embryos remain in cryostorage;

---

[53] http://www.vatican.va/roman_curia/congregations/cfaith/documents/rc_con_cfaith_doc_20081212_sintesi-dignitas-personae_en.html.

[54] Emphasis added. Cindy Wooden, "Adopting Embryos Raises Moral Questions, Vatican Officials Say," *Catholic News Service*, 12/12/2008.

[55] http://www.usccb.org/issues-and-action/marriage-and-family/natural-family-planning/resources/upload/Winter-spring2009.pdf.

23,480 were discarded after being taken out of cryostorage; and about 5,900 were set aside for scientific research[56].

2015 - On February 24, the United Kingdom becomes the first country in the world to allow creating "3-parent" IVF babies. The IVF technique involves creating a genetically modified human embryo by combining: 1) the nuclear DNA (nDNA) from the mother; 2) the mitochondrial DNA (mDNA) from a female donor's egg; and 3) fusing the genetically modified ovum with a sperm from the father using *in vitro* fertilization. Critics fear it will lead to creating "designer babies." "Where will it lead?" said Conservative lawmaker Fiona Bruce during the debate, "The answer has to be that we stop here. The answer has to be that we say this is a red line in our country, as in every other country in the world, that we will not cross"[57].

2015 – On June 17, the U.S. Congress introduces a bill to prevent the FDA [U.S. Food and Drug Administration] from using federal funds for research that involves creating viable embryos

---

[56] https://www.telegraph.co.uk/news/healthnews/9772233/1.7-million-human-embryos-created-for-ivf-thrown-away.

[57] https://www.scientificamerican.com/article/britain-votes-toallow-world-s-first-3-parent-ivf-babies/?print=true.

with heritable genetic modifications, or genetically modifying human sperm or eggs to create such an embryo[58].

2015 – On September 15, the European Court of Human Rights, in a surprise ruling, upholds Italy's 2004 ban on destructive embryonic research and "affirmed Italy's right to protect human embryos as it sees fit"[59].

2015 – On December 22, the Thomas More Society files an amicus brief in two court cases, i.e., *McQueen v. Gadberry* (in Missouri) and *Loeb v. Vergara* (in California), that cites the irrefutable scientific proof that human life begins at fertilization. The amicus brief states that the "current, not antiquated, scientific understanding regarding human embryonic development should drive the Court's ruling. As a matter of scientific fact, the frozen embryos are not potential human life, but actual living human beings. Their 'potential' is for further development consistent with the embryo's body plan that came into existence at the time of the binding of sperm and egg membrane…. Human embryos may not be treated as property and are not merely 'entities deserving of

---

[58] https://nature.com/news/us-congress-moves-to-block-human-embryo-editing.

[59] https://www.lifesitenews.com/news/in-surprise-ruling-european-court-of-human-rights-allows-italys-ban-on-embryo-research.

special respect.' They are human beings and should enjoy equal protection of the laws as persons"[60].

2016 – On February 5, James Clapper, the U.S. Director of National Intelligence, added "gene editing" to the official list of threats that are considered "weapons of mass destruction."

2016 – On April 26, research scientists at Northwestern University in Chicago observe for the first time in human history the "transformative" moment of human conception. The paper is titled, "The Zinc Spark is an Inorganic Signature of Human Egg Activation." Sperm enzyme was used to activate human eggs and their "activation" was made visible using a fluorescent dye to view the release of billions of zinc ions that are expelled from the human ovum at the instant of egg activation, i.e., human conception.

2016 – On September 27, *New Scientist* publishes an article, "Exclusive: World's first baby born with new '3-parent' technique." The method used was approved in the UK. It is called pronuclear transfer and "involves fertilising both the mother's egg and a donor egg with the father's sperm. Before the fertilised eggs start dividing into early-stage embryos, each nucleus is removed. The nucleus from the donor's fertilised egg

---

[60] https://www.thomasmoresociety.org/thomas-more-society-asserts-scientific-fact-embryos-are-human-not-property.

is discarded and replaced by that from the mother's fertilised egg"[61].

2016 – On November 7, a conference at the Center for Bioethics at Harvard Medical School on the "Ethics of Early Embryo Research" proposes to *extend* the "14-Day Rule."

2016 – On November 23, the Human Fertilisation and Embryology Authority (HFEA) reports to Parliament that since 1990, a total of 1,687,260 human embryos were transferred to uteri and a total of "2,315,262 human embryos were deliberately discarded during IVF treatment" in the United Kingdom[62].

2017 – On January 7, research scientists publish an article titled, "Zinc sparks induce physiochemical changes in the egg zona pellucida that prevent polyspermy" in *Integrative Biology*. (N.B. The experiment observed fertilization in mice.)

2017 – "Embryo Adoption Before and After *Dignitas personae*: Defending an Argument of Limited Permissibility" by Sarah-Vaughan Brakman and Darlene Fozard Weaver is published by Springer International Publishing in

---

[61] https://www.newscientist.com/article/2107219-exclusive-worlds-first-baby-born-with-new-3-parent-technique/.

[62] https://catholicherald.co.uk/2016/11/23/almost-2-5-million-human-embryos-destroyed-ivf-in-britain-since-1990/.

Contemporary Controversies in Catholic Bioethics. They argue that Embryo Adoption (EA) "is not excluded in *DP* and that it is grounded in the moral and social teaching of the Church." It is the only moral option for spare frozen embryos[63].

2017 – On November 24, a baby girl was born who had been frozen for 24 years. Her adoptive parents were 26 years old. Emma Wren Gibson was frozen as an IVF embryo in 1992.

2018 – On March 3, a cryostorage tank containing 4,000 frozen human embryos and ova malfunctioned causing the death of all of the embryos in Cleveland, Ohio. On the same day, a second cryostorage tank failed in San Francisco, destroying at least another 2,000 frozen embryos. Class-action lawsuits are being filed by hundreds of affected patients. Many are suing for the "wrongful death" of their "children."

2018 – On November 26, a Chinese scientist announced that he had defied international norms and genetically edited the human genome of two baby girls using CRISPR-Cas 9 to disable a gene that would make the babies resistant to infection with H.I.V. Scientists, ethicists, and policymakers around the world have reacted with alarm. The long-feared genetic engineering of human embryos had become a reality.

---

[63] https://link.springer.com/chapter/10.1007/978-3-319-55766-3_12.

2019 – On January 16, U.S. Senator Rand Paul introduces the Life at Conception Act "to implement equal protection under the 14th Amendment to the Constitution of the United States for the right to life of each born and preborn human person"[64]. This Act "declares that the right to life guaranteed by the Constitution is vested in each human being" and that the "terms 'human person' and 'human being' include each member of the species homo sapiens at all stages of life, including the moment of fertilization or cloning, or other moment at which an individual member of the human species comes into being." In the words of Senator Rand Paul, "it is time for Congress to recognize the right to life is guaranteed to all Americans in the Declaration of Independence, and it is the constitutional [duty] of Congress to ensure this belief is upheld. (…) The Life at Conception Act legislatively declares what most Americans believe and what science has long known – that human life begins at conception, and therefore is entitled to legal protection from that point forward"[65].

2019 - On March 13, the world's leading CRISPR scientists, including its two discoverers, published an article in Nature, a

---

[64] Senator Rand Paul's quote was taken from his press release: https://www.paul.senate.gov/news/sen-rand-paul-introduces-life-conception-act.

[65] https://www.congress.gov/bill/116th-congress/senate-bill/159/text.

British journal considered by many to be the world's leading multidisciplinary science journal, calling "for a global moratorium on all clinical uses of human germline editing — that is, changing heritable DNA (in sperm, eggs or embryos) to make genetically modified children." The scientists also recommend that "there should be a fixed period during which no clinical uses of germline editing whatsoever are allowed. As well as allowing for discussions about the technical, scientific, medical, societal, ethical and moral issues that must be considered before germline editing is permitted, this period would provide time to establish an international framework." Finally, they recommend that while many nations "might choose to continue the moratorium indefinitely or implement a permanent ban" other nations "might well choose different paths, but they would agree to proceed openly and with due respect to the opinions of humankind on an issue that will ultimately affect the entire species"[66].

2019 – On May 2, the 8th District Ohio Court of Appeals rules that frozen embryos are not persons. The 2-1 decision states that "the pre-implanted embryos being stored at University Health System facilities were not capable of independent survival." Bruce Taubman, the attorney for the families who lost their frozen embryos, says he will appeal the appellate

---

[66] Emphasis added to the citation from: https://www.nature.com/articles/d41586-019-00726-5.

court's decision to the Ohio Supreme Court because it is a scientific fact that "embryos are human beings"[67].

2019 - On May 15, Alabama becomes the first U.S. state to pass a bill that bans abortion even in cases of rape and incest. The bill was signed into law by Governor Kay Ellen Ivey, Alabama's first female Republican governor, who said, "this legislation stands as a powerful testament to Alabamian's deeply held belief that every life is precious and that every life is a sacred gift from God." The new law could set up "a court fight that Republicans hope will end with the Supreme Court overturning *Roe v. Wade*"[68].

2019 – By Divine Providence, it is at this critical moment in human history that *Conception: An Icon of the Beginning* by Francis Etheredge is published as an invaluable resource for every concerned citizen, scientist, bioethicist, philosopher, theologian, priest, religious, doctor, lawyer, statesman, and politician. It uniquely brings together, into a single volume, the collective wisdom found in philosophy, theology, science, and the law that defines and defends the beginning of human life at the first instant of conception.

---

[67] https://www.cleveland.com/news/2019/05/universityhospitals-fetility-case-appellate-court-rules-lost-embryoswere-not-living-person.html.

[68] https://politico.com/story/2019/05/15/alabama-governor-abortion-1327389.

The purpose of looking back at the past 200 years – since the scientific discovery of human fertilization – is to help explain the major historical importance of *Conception: An Icon of the Beginning* written and compiled by Francis Etheredge. The human race is at an existential crossroads and a fierce battle is being waged worldwide between a Culture of Life built upon the "laws of nature and of Nature's God" versus a culture of death that is attempting to defy these God-given laws. Human conception is the very heart of this battle.

Conception: An Icon of the Beginning should be read cover to cover in order to fully understand the urgency for coming together as the children of God that we truly are, made in the very image and likeness of God who so loved us that He became one of us at the first instant of His own conception, in order to reveal to us the great mystery of man.

A LOOK "*IN THE BOOK*":
Five Points from Chapter Five

"But as regards the book, only refer to what is in the book so that you can go forward." ~ Francis Etheredge

It would be impossible to list all of the significant and truly visionary contributions that are contained within this interdisciplinary masterpiece by Francis Etheredge that is dedicated to exploring and revealing in greater depth the profound natural and supernatural truths that are have been hidden for so many centuries deep within the first instant of human conception. But five contributions that Etheredge examines in his Fifth Chapter are of such great importance for all humanity that they deserve to be *referred to* – and briefly quoted – in order to *go forward* and conclude this End Word. These five exceptionally significant contributions are best presented using direct quotes from Francis Etheredge himself.

Conception – like Creation – is an Irrevocable Beginning

"In other words, the mystery of the beginning of a human life *recapitulates* the profundity of the beginning of creation which began with the act of God at the beginning of everything."

"[T]he moment of conception is precisely this: that it is an irrevocable beginning: a beginning that does traverse time: that unfolds through time 'without end". Thus the 'moment' of that beginning is complete in the very moment it comes to exist."

"[T]he moment of the sperm and the egg's union which, in an expression of the will of the Creator, is when God *repeats anew His creation from nothing, and completes His creation of each one of us.*"

"Therefore the marvellous fact of an observable beginning to human life *is the beginning which is also and necessarily the data of philosophy and theology* – because this also an act of God as radical as the beginning to which the Book of Genesis directs us."

The Conception of Mary and the Incarnation of Christ

"[T]here is a kind of "Patristic Principle' based upon a number of texts of the Fathers of the Church which, in effect, lead to the conclusion that whatever is true of human conception is redeemed by the coming of Christ; and, therefore, as the nature of human conception becomes better understood, so it is clearer that redemption begins with the beginning of the Conception of Mary and the Incarnation of Christ."

"God's conception of Mary is an ensouling action of God which takes up the wholly natural contribution of her parents into the history of salvation."

"[W]hat can be recognized as true of human conception will apply to Mary's *Immaculate Conception*"… and "what follows from her graced conception assists us to understand the first instant of her conception and, therefore, the first instant of our conception."

"Moreover, given that a personal grace required a personal subject, it would follow that Mary was 'present' from the first instant of her conception; and in view of what we know of human conception, the first instant of human conception is the formation of the embryonic wall following the sperm's penetration of the ovum."

"Mary, then, after Adam and Eve, is the pre-eminent case of the human person united to the Son of God; and, in so being, it makes radical sense that she is wholly without sin and completely human, one in soul and body, from the first instant of her conception."

"In the case of the Incarnation of the Son of God (…) the very conception of Christ entails a first moment of the fertilization of the human ovum as it is animated by Christ's human soul and, therefore, begins an unfolding development which expresses the personhood of the Christ-child."

"In other words, just as the conception of the Virgin Mary entailed a first instant from which 'she' was conceived, one in body and soul, so there is a first instant from which Christ was miraculously conceived, one in body and soul."

"If then Christ is God from God uninterruptedly, then it follows that Christ is man from Mary in the first instant of 'becoming flesh' (Jn 1:14)."

The Sperm-Inclusive-Enclosing of the Embryonic Wall

"What can be reasonably established about human conception is that there is an obvious start to the new entity of the human embryo: the formation of the embryonic wall on the penetration of the ovum by the sperm."

"[I]n the 'sperm-inclusive-enclosing' of the embryonic wall, there is a natural 'sign' capable of expressing an 'inner mystery': the natural outward sign of the enclosing embryonic wall expressing the inner moment in which God determines there to be an animating human soul."

"Thus we can argue that God acts in a way which gives witness to His action, not because He needs it but because it is part of the Creator's communication to us of the nature of human being and the mystery of God."

"*Orta a fusione*" in the Latin Text of *Donum Vitae*

"[T]he Latin text of *Donum vitae* simply says that the zygote comes to exist "orta a fusione" (arising from a fusion).

"The Latin text (…) does not include the defining note which says, 'The zygote is the cell produced when the nuclei of the two gametes have fused.'"

"Thus the Latin text could mean that the zygote comes into existence from the first instant of the fusion of the sperm and the egg which, as we know, developmentally unfolds uninterruptedly from then on."

The Morality of Embryo Transfer and Adoption

"At its simplest, a child conceived outside the womb is without the immediate possibility of benefitting from the mother's nurturing contribution to the completion of embryological development; and, therefore, there is a natural injustice to the child, who in 'his' own way, cries out for redress: a voiceless cry which appeals to us the more it is almost inaudible in being submerged in freezing temperatures."

"[T]he sophistication of the methods of preserving the child in storage is an indication of the lengths to which 'we' will go to 'hide' the humanity of the frozen child."

"[The] procedure of transferring the embryo is radically different from the artificial methods of conception" … "The reason that the transfer of a child to the womb of a woman is not to be confused or assimilated to *in vitro* fertilization and its methods is that this child now exists; and, intrinsic to his or her existence, is the natural right to completing human development which, in the case of the embryonic stage of a child's development, requires the nurturing presence of a woman willing to be an adopting mother."

"The Church has already indicated that embryo transfer is acceptable when it is for the good of the embryo and, therefore, the possibility of embryo adoption is also acceptable for the same reason: it is necessary for the good of the homeless embryonic child."

A LOOK "FORWARD":
Starting from the awe-inspiring and "transformative" first moment of human conception

In order to *look forward* to a new beginning I can think of no better starting point than by studying the awe-inspiring and "transformative" first moment of human conception – which is the primary subject of this entire book – which became visible for the world to see, for the first time in history, on August 27, 2016.

"Seeing Life in a New Light: Scientists Capture 'Transformative' Moment of Human Conception"

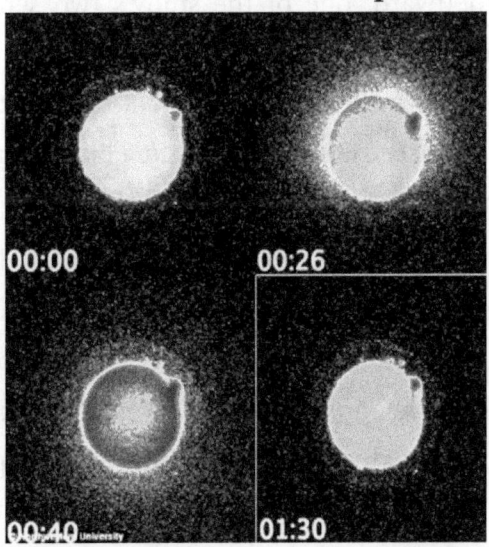

(Photo and Article Published in Northwestern University News on 4/27/2016: https://on.rt.com/7b5z)

On April 26, 2016, *Scientific Reports* published, "The Zinc Spark is an Inorganic Signature of Human Egg Activation" that documents the transformative moment of human conception and also explains the scientific significance of egg activation which is caused when the sperm and egg first fuse.

> "Egg activation refers to events required for transition of a gamete into an embryo, including establishment of the polyspermy block, completion of meiosis, entry into mitosis, selective recruitment and degradation of maternal mRNA, and pronuclear development. [...] Together, these results demonstrate critical functions for zinc dynamics and establish the zinc spark as an extracellular marker of early human development"[69].

Teresa Woodward, one of the study's two senior authors at Northwestern, commented, "All biology starts at the time of fertilization, yet we know next to nothing about the events that occur in the human [...] to see zinc radiate out in a burst from each human egg was breathtaking"[70].

Let us now carefully observe each of the four photographs above which capture the image of the *human egg* moments before human conception occurs, and the two photographs that capture the image of what would be the *human embryo* immediately *after* the 'transformative' moment of human conception when the sperm-activated

---

[69] https://www.nature.com/scientificreports/articles/srep24737.
[70] https://www.rt.com/usa/341063.

human egg has become a living human embryo, a living human being; i.e., a person.

In the first photograph, taken at 00.00 seconds, a sperm's enzyme has entered the human ovum, but fertilization has not yet occurred. The next photograph, however, which is taken at 26 seconds, reveals the "transformative" moment of human conception as billions of zinc ions, referred to as "zinc sparks," suddenly burst through the cellular wall of the newly fertilized human egg that has become a human embryo.

This powerful expulsion of billions of excess zinc ions triggers the "polyspermy block" constituting an independent act in which the newly conceived human being protects itself within a hardened and impenetrable membrane.

The two gametes, one egg and one sperm, have fused with each other and have been transformed into a single living *organism,* a living *one*-cell human embryo, who is now a unique human being, completely distinct from its parents, with its own pronuclei, and who is already a male or a female.

This earliest stage of the human embryo is scientifically known as an "ootid" and the new human being has already begun its own self-directed process of fusing together its own two *pronuclei* into a single nucleus. When the single-cell human *ootid* completes the fusion of its two pronuclei, approximately 24 hours later, the *ootid* will then be known as a single-celled human *zygote* with a completely fused nucleus.

In the next two photographs, taken respectively at 40 seconds and at 01:30 minutes, the "zinc sparks" are clearly diminishing because *fertilization* has taken place and the outer membrane of the new human embryo has immediately hardened to protect itself and its integrity from polyspermy.

The following diagram shows the moment before and the moment after the transformative instant of fertilization:

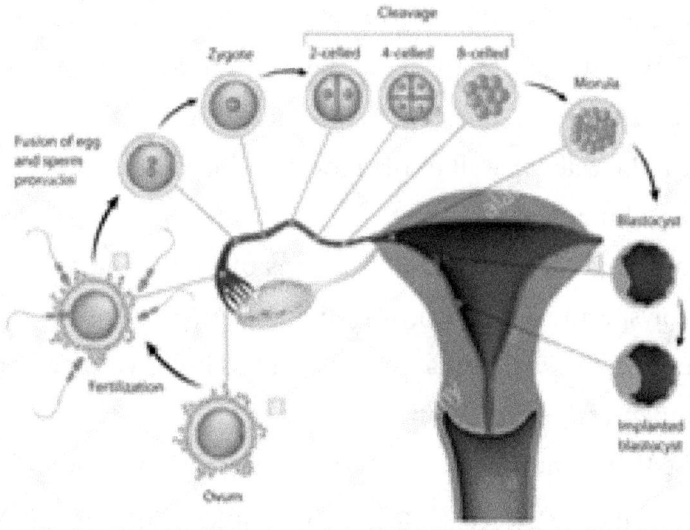

In particular, the third and fourth photographs taken immediately after fertilization capture the amazing first images of what each one of us looks like immediately after conception. As the diagram above also clearly shows, as newly conceived human embryos, we were already progressing toward our *second observable act*, namely, the *self-directed* fusion of our two haploid pronuclei into a single nucleus with the unique set of 46 chromosomes that each one of us possesses, as *ootids*, from the very first instant of fertilization. From that first instant onward, each one of us will continue to possess those very same 46 chromosomes during our entire earthly lives and, after our death and resurrection from the dead, for all eternity.

How do we know this? Because faith and reason, along with science and theology, must necessarily lead to the same ultimate truths regarding the mystery of God and the mystery of mankind. These new, breathtaking and irrefutable scientific facts can and should help every one of us to better understand what God has so magnificently revealed to all humanity through the magisterially defined dogmas of the Immaculate Conception of Mary and the Incarnation of the Second Person of the Blessed Trinity, Our Lord and Savior, Jesus Christ.

What can these absolutely stunning photographs reveal to us about Mary's Conception and Jesus' Incarnation, and about every human being who has ever been conceived?

How wonderful it is that, once again, faith and reason so perfectly support each other. In 1854, in *Ineffabilis Deus,* the Magisterium teaches as divinely revealed that Mary's "soul, in the first instant of its creation and in the first instant of the soul's infusion into the body, was, by a special grace and privilege of God, in view of the merits of Jesus Christ, her Son and the Redeemer of the human race, preserved free from all stain of Original Sin." Today, science has made it easier for humanity to understand and accept the truth of this dogma which must have occurred at the transformative moment of Mary's conception when her father's sperm first activated her mother's egg. As loving spouses, Mary's parents were predestined from all eternity to co-create their daughter, Mary, together with the Blessed Trinity: God the Father who created and infused her spiritual soul; God the Son who redeemed her; and God the Holy Spirit who sanctified her at the first instant of her Conception.

To Mary "did the Father will to give His only-begotten Son – the Son, Whom, equal to the Father and begotten by Him, the Father loves from His Heart – and to give this Son in such a way that he would be the one and same common Son of God the Father and of the Blessed Virgin Mary. It was she whom the Son Himself chose to make His Mother and it was from her that the Holy Spirit willed and brought it about that He should be conceived and born from whom He Himself proceeds"[71].

Beautifully implicit, therefore, in the Dogma of the Immaculate Conception of the Blessed Virgin Mary, is the necessary truth that God must also immediately create and directly infuse our spiritual and rational souls into our one-cell human bodies at the very first instant of human conception. To deny that this is true for each human being would be tantamount to denying that it is true for Mary who is as fully human as we are, except for sin.

When these beautiful pictures that capture the transformative moment of human conception were first made public, it was not surprising that many commentators hailed them as a tiny glimpse, as a miniature "icon," of the very beginning of Creation itself, when God the Creator summoned the entire universe into existence from nothing, *ex nihilo*, with His almighty command, "Let there be light!"

In Mary, we clearly can see that her human conception was truly an "icon" of the beginning because from "the very beginning, and before time began, the eternal Father chose and prepared for His only-begotten Son a Mother in whom the Son of God would become

---

[71] http://www.papalencyclicals.net/pius09/p9ineff.htm.

incarnate and from whom, in the blessed fullness of time, He would be born into this world."

Likewise, each of us has been known and loved by God from all eternity, since before the beginning of the world. We, too, like Mary and Jesus, have divinely created spiritual souls made in the image and likeness of God that are immediately created and directly infused in our wondrous, one-cell human bodies at the first moment of our human conception. "In our bodies," Saint John Paul II once said, "we are mere specks in the vast created universe, but by virtue of our souls we transcend the whole material world"[72].

Conclusion: A Bold New Beginning

It is my sincere hope that this End Word has served to highlight the historical importance of this great work as we continue together to forge a bold new beginning for humanity.

May "A Look *Back*" provide helpful information to better understand and confront our present day challenges and inspire us to research, write, and actively pursue important legislative efforts such as Germany's Embryo Protection Act, Italy's Medically Assisted Reproduction Law, and America's ongoing effort to pass a Life at Conception Act.

---

[72] The quotation from St. John Paul II was found in the following book: *Lessons for Living*, edited by Joseph Durepos, Loyola Press, Chicago, Illinois, 2004, p. 47.

May "A Look *in the Book*" inspire us to ponder and further investigate so many of the profoundly significant and visionary contributions that Francis Etheredge has proposed.

May "A Look *Forward*" encourage us to work diligently in every professional field and academic discipline. There is an urgent need to promote a greater international clarity and consensus that defends human life at conception. Science must be required to ethically regulate itself, and strict laws must be enacted to protect vulnerable human life from the first moment of conception until natural death. Within the Church, given the global assault and destruction of human lives before birth, and given the now irrefutable scientific evidence that human life begins at conception, there is a great need for a definitive formulation regarding God's immediate creation and infusion of the spiritual soul at conception. Following the example of Francis Etheredge himself, let us invite people of goodwill from every religious tradition to collaborate in a vigorous defense of human life at conception.

Finally, on behalf of all of his contributors, I want to thank Francis Etheredge for inviting us to join him in this exceptional work and serving as a great inspiration to us all.

May these final words which were spoken on the Feast of the Presentation of the Lord by our beloved Holy Father, Saint John Paul II, to whom Francis Etheredge has dedicated this entire book, express in some small way our gratitude.

The Light of Life

"Today the Church blesses the candles which give light. These candles are, at the same time, a symbol of the other light, the light of Christ. He began to be light from the moment of his birth. He was revealed as light to the eyes of Simeon on the fortieth day after his birth. Then he remained as light for thirty years in the hidden life of Nazareth. Subsequently, he began to teach.... He said: *I am the light of the world: he who follows me will not walk in darkness, but will have the light of life* (Jn 8:12). When he was *crucified there was darkness over all the land* (Mt 27:45) but on the third day this darkness made way for the light of the Resurrection.

"The light is with us! What does it illumine? It illumines the darkness of human souls. The darkness of existence. *Man makes a perennial and immense effort to open up a way and arrive at light; the light of knowledge and existence. How many years does not man at times dedicate to clarifying some fact for himself, to finding the answer to a given question!* And how much personal toil it costs each one of us in order that – through everything in us that is "dark," shadowy through our "worse self," through the man subjugated by the lust of the flesh, the lust of the eyes, and the pride of life (cf. 1 Jn 2:16) – we can reveal what is luminous: *the man of simplicity, of humility, of love, of disinterested sacrifice; the new horizons of thought, of the heart, of will, of character. The darkness is passing away and the true light is already shining* (1 Jn 2:8).

"If we ask what is illumined by this light, recognized by Simeon in the Child forty days old, the answer is as follows… It is the answer to your life"[73].

---

[73] Emphasis added. The quotation comes from the Magnificat, Vol. 20, No. 12, February 2nd, 2019, "Feast of the Presentation of the Lord, Meditation of the Day, St. John Paul II, pp. 45-46.

Essay V

Journal of Bioethics in Law & Culture
Summer 2024 ● vol. 7 issue 3, pp. 119-134

Understanding the Importance of the Alabama Supreme Court Decision: Frozen IVF Embryos are Legally Protected "Extrauterine" Unborn Children

Elizabeth Bothamley Rex

https://www.societyofstsebastian.org/summer2024-bioethics-law-cul

THE LANDMARK ALABAMA SUPREME COURT DECISION

On February 16, 2024, in a 7-2 decision, the Alabama Supreme Court issued a first-in-the-nation ruling that extrauterine IVF human embryos are "unborn children" under Alabama's Wrongful Death of a Minor Act regardless of their location, either inside or outside of the womb.

The Alabama Supreme Court Decision involved a lawsuit brought by three couples whose frozen embryos were negligently destroyed in 2020 at a fertility clinic by an unauthorized intruder who entered the cryostorage facility through an unlocked door, removed several frozen embryos from a cryostorage container, and then dropped them on the floor, killing all of them. (1)

The distraught parents sued the Mobile Medical Center for the tragic and negligent death of their IVF children "under Alabama's Wrongful Death of a Minor Act. (1) "A lower court ruled in favor of the clinic, arguing that the embryos did not meet the criteria for 'person' or 'child' (…) but, on appeal, the Supreme Court overturned the lower court decision and ruled in favor of that the law 'applies to all children, born and unborn, without limitation.'" (2)

In the majority ruling, Justice Jay Mitchell officially presented the Supreme Court's landmark decision:

"This Court has long held that unborn children are 'children' for purposes of Alabama's Wrongful Death of a Minor Act, (…) a statute that allows parents of a deceased child to recover punitive damages for their child's death. The central question (…) is whether the Act contains an unwritten exception to that rule for extrauterine children – that is, unborn children who are located outside of a biological uterus at the time they are killed. Under existing black-letter law, the answer to that question is no: the Wrongful Death of a Minor Act applies to all unborn children, regardless of their location." (3)

In his long and detailed majority opinion, Justice Jay Mitchell clearly and historically ruled:

"Unborn children are 'children' under the Act, without exception based on developmental stage, physical location, or any other ancillary characteristics. (…) It applies to all children, born and unborn without limitation." (4)

Following Justice Mitchell's main opinion, Chief Justice Tom Parker added his 22-page concurring opinion explaining in great legal detail how this decision was solidly based upon the "Sanctity of Unborn Life Amendment" to the Alabama Constitution that was voted into law by a majority of the People of Alabama back on November 6, 2018:

"A good judge follows the Constitution instead of policy, except when the Constitution itself commands the judge to follow a certain policy. In these cases, that means upholding the sanctity of unborn life, including unborn life that exists outside the womb. Our state Constitution contains the following declaration of public policy: 'This state acknowledges, declares, and affirms that it is the public policy of this state to recognize and support the sanctity of unborn life and the rights of unborn children, including the right to life.'" (5)

"This case presents an opportunity for this court to continue a line of decisions affirming Alabama's recognition of the sanctity of life from the earliest stages of development. We have done so in three recent cases (…); we do so again today." (6)

"Accordingly, any legislative (or executive) act that contravenes the sanctity of unborn life is potentially subject to a constitutional challenge under the Alabama Constitution." (7)

Importantly, in his concurring opinion Chief Justice Parker addressed the concerns raised by the lone dissenting judge, Justice Cook, who had argued that the application of the Sanctity of Life Act and the Wrongful Death of a Minor Act to frozen embryos "[would] have disastrous consequences for the in vitro fertilization ("IVF") industry in Alabama." (8) The Chief Justice acknowledged Justice Cook's prescient concerns but correctly observed:

"[I]t is for the Legislature to decide how to address this issue. I note briefly that many other Westernized countries have adopted IVF practices or regulations that allow IVF to continue while drastically reducing the chances of embryos being killed, whether in the creation process, the implantation process, the freezing process, or by willful killing when they become inconvenient. For decades, IVF has been largely unregulated in the United States, with some commentators even comparing it to the Wild West. (…) (9)

"In Alabama, the only statutes that mention IVF address the issue of determining parentage of children conceived through IVF, but they do not govern the practice of IVF itself. (…) (10)

"If the Legislature agrees that it is time to regulate the IVF industry, then the good news is it need not reinvent the wheel. Other Westernized countries have given Alabama some examples to consider." (11)

PUBLIC REACTION & PRO-LIFE SUPPORT

Much like the overwhelmingly joyous news regarding the United States Supreme Court decision in the Dobbs case that had finally overturned the infamous Roe v. Wade decision on June 24, 2022, so, too, the recent Alabama Supreme Court decision on February 16, 2024, was immediately celebrated as a landmark moment in the history of our Nation to secure the Right to Life for all unborn children, including roughly 1.5 million frozen, unborn IVF children here in the United States.

The Alabama Supreme Court's historic decision was immediately featured by the mainstream media as "breaking news" and rapidly spread across the nation and around the world. The *National Catholic Register* reported that "President Biden denounced the decision as 'outrageous and unacceptable'" The same NCR article also reported that former-President Trump immediately called on Alabama's state legislators to "ensure access to IVF, saying his party 'should always be on the side of life and on the side of mothers and fathers and beautiful little babies." (12)

Praise and public support for Alabama's Supreme Court's decision also began pouring in from conservative media and the leaders of prominent pro-life and Catholic organizations.

According to Danielle Pimentel, a policy attorney at Americans United for Life, the Court's decision was "focused on Alabama law and will stay within Alabama." It is Pimentel's opinion that the ruling "doesn't limit IVF or access to it. It simply ensures that both the parents and the children are protected under the Wrongful Death of a Minor Act" which will allow the parents to bring a civil claim against a fertility clinic if it is proven to have acted negligently. (13)

Katie Daniel, the State Policy Director at SBA Pro-Life America, said in a statement to Catholic News Agency that the court in its ruling "recognized what is obvious and a scientific fact – life begins at conception." (14)

Lila Rose, Founder and President of Live Action, called the decision "an important step towards applying equal protection for all." (15)

The Catholic Medical Association (CMA) issued a Press Release to its 2,600 members and health care professionals who belong to 119 local guilds praising the Alabama Supreme Court "for its recent decision recognizing the biological fact that an embryo is a human being from the moment of fertilization, upholding the state's recognition of the right to life for every human being from the moment of conception." In addition, Marie Hilliard, J.C.L., Ph.D., RN, the co-chair of CMA's Ethics Committee added, "The Supreme Court of Alabama's ruling not only protects the unborn but also assures parents that their precious embryos have the protection of the law." (16)

Elizabeth Kirk, co-director of the Center for Law & the Human Person at The Catholic University of America Columbus School of Law in Washinton, commented on the historic fact that while the Alabama Supreme Court "had previously held that unborn children in the womb are 'children,' in this ruling, however, "the court held that the word 'child; in the statute includes unborn children regardless of location, whether in or outside of a biological womb." Kirk added, "all of us should welcome laws and court decisions that comport with the truth of the human person, including the dignity of all human life from conception to natural death." (17)

And, the President of the National Catholic Bioethics Center, Dr. Joseph Meaney, when asked on February 23rd to comment on

Alabama's recent court ruling during his interview with "EWTN News in Depth," supported the Court's landmark decision stating:

> "We become new human beings at the moment of conception. The Church is very clear about this, and science is very clear about this. We have to realize that if life begins at conception, then all those conceived human beings should be protected. Whether they're in an IVF lab or in the wombs of their mothers, these are new human beings that deserve protection." (18)

CATHOLIC AND PRO-LIFE CONCERNS VERSUS ALABAMA'S "TEMPORARY FIX"

Over the next few weeks, however, the great news of this historic court decision was quickly replaced by the overwhelmingly chaotic news that three of Alabama's largest fertility centers were "pausing" their IVF treatments out of fear of being criminally or civilly prosecuted for procedures that were previously considered as "standard" IVF protocol.

Lawmakers in both parties began to work at "warp speed" to draft legislative bills in the Alabama House and Senate to address the growing crisis and large protests from hundreds of IVF patients affected by the sudden "pausing" of the IVF clinics. On February 22, 2024, just 6 days after the Supreme Court's decision, Democrat Representatives were the first to introduce HB 225, a bill that would completely dehumanize an unborn child located outside of the uterus.

> "Existing law protects the rights of unborn children in certain circumstances. The Alabama Supreme Court has held that "extrauterine children: that is, unborn children "located outside of a biological uterus," are considered unborn children for purposes of the Wrongful Death of a Minor Act.
>
> "This bill would provide that any fertilized human egg or human embryo that exists outside of a human uterus *is not considered an unborn child or a human being for any purpose under state law.*" (Emphasis added) (19)

On February 27, 2024, Republican Representatives and Senators introduced two similar bills, HB 237 and SB 159 which stated:

> "Relating to in vitro fertilization: to provide civil and criminal immunity for death or damage to an embryo to any individual or entity when providing or receiving services related to in vitro fertilization." (20)

On Thursday, February 29, following hours of debate in both chambers, a vote was taken to proceed: HB 159, sponsored by Rep. Terri Colins, R-Decatur, passed the House on a 94-6 vote, and SB 159, sponsored by Sen. Tim Melson, R-Florence, passed the Senate on a 34-0 vote.

Meanwhile, in the Democrat-controlled US Senate, another legislative war erupted over an attempt to federally protect access to IVF as a direct response to the Alabama Supreme Court decision. On February 27, 2024, Sen. Tammy Ducksworth (D-IL) introduced a bicameral bill (S. 3612) called the *Access to Family Building Act* (21) "to establish a federally protected right to IVF access, preempting state-imposed restrictions." (22) On February 28, the USCCB officially released a Letter (23) that was addressed to every US Senator "urging lawmakers to oppose [the] bill that would create a federally sanctioned right to access in vitro fertilization (IVF)-a fertility treatment that has resulted in the deaths of millions of human embryos in the United States." (24)

The Catholic bishops expressed their strong opposition to in vitro fertilization which "involves the creation of countless preborn children and results in most of them being frozen or discarded and destroyed." (24) The bishops strongly opposed the terms of the bill that could be "interpreted to fabricate and impose new rights to human cloning, gene editing, making human-animal chimeras, reproducing children of a parent who is long deceased, engaging in the buying and selling of human embryos, commercial gestational surrogacy, and more." In addition, they warned the senators that this bill would require faith-based organizations to provide insurance coverage for IVF in their employee health plans and thus "would be the first law ever to exempt itself from the longstanding Religious Freedom Restoration Act (NRFR) ... passed in the Senate by a vote of 97 to 3 in 1993." (25)

Fortunately, thanks to an "objection for the Senate floor from Sen. Cindy Hyde-Smith, R-Mississippi, the bill was blocked from advancing via unanimous consent and must go through the committee process before it can receive a vote." (26)

On the same day that the USCCB letter to the Senators was written, March 4, 2024, another important letter addressed to Governor Kay Ivey of Alabama was publicly released. It was titled, the *Pro-Life Coalition Letter Opposing Alabama IVF Legislation* and was signed by over a dozen pro-life leaders at prominent pro-life organizations who are "united in opposition to the passage of current legislation headed to [her] desk (SB159/SB 237)" and encouraging the Governor and "the elected officials of Alabama to slow down and study the ethical implications of this highly complex topic." (27) The Pro-Life Coalition Letter concluded by stating:

> "Any political determination that takes up the question of how we treat and protect human lives – no matter how young – must resist an ideology that treats human beings as expendable commodities. Any legislation on this issue must take into consideration the millions of human lives who face the fate of either being discarded or frozen indefinitely, violating the inherent dignity they possess by virtue of being human" and they called on Governor Ivey "to lead with courage on an issue that holds such moral gravity." (28)

Three days later, on in the evening of Wednesday, March 7, Governor Kay Ivey signed the bill that had just been passed by the Alabama

Legislature with an overwhelming, bi-partisan majority: in the House the vote was 81-12, and in the Senate the vote was 29-1. After signing the bill into law, Governor Ivey immediately released the following statement on X:

> "The overwhelming support of SB 159 from the Alabama Legislature proves what we have been saying: Alabama works to foster a culture of life, and that certainly includes IVF. I am pleased to sign this important, short-term measure into law so that couples in Alabama hoping and praying to be parents can grow their families through IVF. IVF is a complex issue, no doubt, and I anticipate there will be more work to come, but right now, I am confident that this legislation will provide the assurances our IVF clinics need and will lead them to resume services immediately.
>
> "Make no mistake about it, though, in the coming days, weeks and months, particularly as we are in the heat of a national election, we will hear a lot of political rhetoric around IVF. Let me say clearly: Alabama supports growing families through IVF. From protecting the unborn to supporting IVF, Alabama is proud we are a pro-life, pro-family state." – Governor Kay Ivey (29)

Legislators on both sides of the aisle have repeatedly expressed concerns that "this is a really bad piece of legislation" that will surely invite lawsuits and constitutional challenges. In fact, a second "wrongful death of a minor" lawsuit has just recently been filed against the same Mobile clinic and is already challenging the very immunity that the Alabama legislature has just provided for IVF providers.

Most of the Alabama legislators realized this legislation was just a "temporary fix" – a band-aid -to reopen the IVF clinics and to allow them more time to prepare a better legislative solution as soon as possible. At a press conference called by the House Republican Caucus after the bill's passage, the sponsor of the bill, Rep. Terri Collin, R-Decatur, explained that finding the right solutions will require "a longer conversation than we can have in five days." (30)

THE URGENT NEED FOR "EMBRYO PROTECTION" LAWS IN EVERY STATE

The Alabama Supreme Court is the first state in the nation to rule that frozen human embryos constitute children under state statute, by ruling that the constitutional provisions in the state's 1872 Wrongful Death of a Minor Act and the state's 2018 Sanctity of Life Act "apply to all children, born and unborn, without limitation" stating that it is "especially true where, as here, the people of Alabama have adopted a constitutional amendment directly aimed at stopping courts from excluding 'unborn life' from legal protection." (31)

Chief Justice Parker concluded his legal opinion in this historic court ruling as follows:

"The people of Alabama have declared the public policy of this State to be that unborn life is sacred. We believe that each human being, from the moment of conception, is made in the image of God, created by Him to reflect his likeness. (…) All three branches are subject to a constitutional mandate to treat each unborn human life with reverence. Carving out an exception for the people in this case, small as they are, would be unacceptable to the People of this State, who have required us to treat every human being in accordance with the fear of a holy God who made them in His image. For these reasons, and for the reasons stated in the main opinion, I concur." (32)

The Magisterium of the Catholic Church confirms the truth of Alabama's historic ruling:

"The inalienable rights of the person must be recognized and respected by civil society and the political authority. These human rights depend neither on single individuals nor on parents; nor do they represent a concession made by society and the state; they belong to human nature and are inherent in the person by virtue of the creative act from which the person took his origin. Among such fundamental rights one should mention in this regard every human being's right to life and physical integrity from the moment of conception until natural death."

"The moment a positive law deprives a category of human beings of the protection which civil legislation out to accord them, the state is denying the equality of all before the law. When the state does not place its power at the service of each citizen, and in particular of the more vulnerable, the very foundations of a state based on law are undermined.... As a consequence of the respect and protection which must be ensured for the unborn child from the moment of conception, the law must provide appropriate penal sanctions for every deliberate violation of the child's rights." (33)

---

*Elizabeth Bothamley Rex, PhD, is an Associate Scholar of The Charlotte Lozier Institute, a former Adjunct Professor of Catholic Bioethics at Holy Apostles College & Seminary, and the President of The Donum Vitae Institute.

Endnotes

(1) Supreme Court of Alabama, October Term, 2023-2024, SC-2022-0515, p 5. https://publicportal-api.alappeals.gov/courts/68f021c4-6a44-4735-9a76-5360b2e8af13/cms/case/343D203A-B13D-463A-8176-C46E3AE4F695/docketentrydocuments/E3D95592-3CBE-4384-AFA6-063D4595AA1D (accessed 2/24/24)

(2) Alanda Rocha, "University of Alabama Birmingham pauses IVF treatments after court ruling," *Alabama Reflector*, February 21, 2024, https://alabamareflector.com/2024/02/21/university-of-alabama-birmingham-pauses-ivf-treatments-after-court-ruling/ (accessed 3/10/2024).

(3) Supreme Court of Alabama, p 2.

(4) Ibid., p 11.

(5) Ibid., p 26.

(6) Ibid., p 33.

(7) Ibid., p 42.

(8) Ibid., p 43.

(9) Ibid., p 43.

(10) Ibid., p 44.

(11) Ibid., p 44.

(12) Richard M. Doerflinger, "Fact and Fiction in the Alabama Embryo Case," *National Catholic Register*, March 10 – 23, 2024, Volume 100, No. 7, p 1.

(13) Daniel Payne/CNA, "After Alabama Supreme Court's Embryo Personhood Ruling, What Comes Next?" *National Catholic Register,* February 24, 2024, https://www.ncregister.com/cna/after-alabama-supreme-court-s-embryo-personhood-ruling-what-comes-next (accessed 2/27/24).

(14) Ibid.

(15) Ibid.

(16) Press Release, "CMA Praises Alabama Supreme Court Decision Recognizing Right to Life of Embryos" *Catholic Medical Association*, February 29, 2024, https://www.cathmed.org/media/press-releases/

(17) Kate Scanlon/OSV News, "Alabama Supreme Court rules frozen embryos are children under wrongful death law," *America Magazine,* February 26, 2024, https://www.americamagazine.org/politics-society/2024/02/22/alabama-supreme-court-ivf-abortion-247362

(18) Peter Pinedo/CNA, "Here's what Trump, Biden, and the Catholic Church are saying about IVF," Catholic News Agency, February 26, 2024, https://www.catholicnewsagency.com/news/256918/here-s-what-trump-biden-and-the-catholic-church-are-saying-about-ivf

(19) Alabama Legislature HB 225 (accessed 3/9/2024) https://www.legislature.state.al.us/pdf/SearchableInstruments/2024RS/HB225-int.pdf

(20) Alabama Legislature HB 159 (accessed 3/9/2024) https://www.legislature.state.al.us/pdf/SearchableInstruments/2024RS/SB159-enr.pdf

(21) Press Release, "Support for Duckworth Bill to Protect IVF Access Grows after Alabama Supreme Court Ruling," March 4, 2024 (accessed 3/11/2024). https://www.duckworth.senate.gov/news/press-releases/support-for-duckworth-bill-to-protect-ivf-access-grows-after-alabama-supreme-court-ruling

(22) EWTN News, *National Catholic Register*, "Catholic Bishops Object to Senate IVF Bill, Warn Against Deaths of Preborn Children," March 1, 2024, (accessed 3/8/2024) https://www.ncregister.com/cna/catholic-bishops-object-to-senate-ivf-bill-warn-against-deaths-of-preborn-children

(23) United States Conference of Catholic Bishops, *Letter to Senators on the Access to Family Building Act, February 28, 2024*. https://www.usccb.org/resources/Letter_Access_to_Family_Building_Act_2024.pdf

(24) EWTN, ibid.

(25) USCCB, ibid.

(26) EWTN, ibid.

(27) Lila Rose et al., Live Action, *Pro-Life Coalition Letter Opposing Alabama IVF Legislation,* March 4, 2024 (accessed 3/10/2024). https://www.liveaction.org/wp-content/uploads/2024/03/Pro-Life-Coalition-Letter-on-Alabama-IVF.pdf

(28) Lila Rose, ibid.

(29) Cf. Steve Ertelt, LifeNews.com, *Alabama Republican Governor Signs Law Protecting IVF,* March 7, 2024. https://www.lifenews.com/2024/03/07/alabamas-republican-governor-signs-law-protecting-ivf/ (accessed 3/7/2024)

(30) Alander Rocha, Alabama Reflector, *Alabama passes law to protect access to IVF treatments,* March 7, 2024 (accessed on 3/7/2024). https://19thnews.org/2024/03/alabama-ivf-bill-legislature-approval/

(31) Cf. Peter Pinedo, Catholic News Agency, Ibid.,

(32) Supreme Court of Alabama, ibid., p 48.

(33) *Catechism of the Catholic Church* n. 2273, citing *Donum vitae* III.

Essay VI

Sebastian's Point / Society of St. Sebastian
Weekly Column - 09 September 2024

Model Laws in Germany, Italy, Louisiana, and Georgia Restrict Both IVF and Cryopreservation to Legally Protect Human Embryos from Further Abuse, Physical Harm, and Death

Elizabeth Bothamley Rex
President, Donum Vitae Institute

https://www.societyofstsebastian.org/ifv-laws-rex

Model Laws in Germany, Italy, Louisiana, and Georgia Restrict Both IVF and Cryopreservation to Legally Protect Human Embryos from Further Abuse, Physical Harm, and Death

The recent landmark Alabama Supreme Court decision on February 16, 2024, that ruled that frozen embryos are "unborn children" regardless of their location or any other ancillary circumstance, has brought much needed attention to the inhumane plight of over one million "unborn IVF children" who are frozen and abandoned in cryostorage tanks in US fertility clinics.

What is not well known to many policymakers and legislators in the United States is that two European countries (Germany and Italy) and two US states (Louisiana and Georgia) have already enacted model legislation to restrict IVF and cryopreservation to protect the lives and welfare of human embryos created in vitro from further abuse, harm, and death.

The focus of this brief article is to encourage its readers to research these important laws in Germany, Italy, Louisiana, and Georgia and to further promote their enactment in all 50 states.

Germany
Germany has strict regulations regarding cryopreservation and experimentation on embryos due to its historical, ethical, and legal considerations.

1) <u>Historical Context</u>: Germany's approach to bioethics is heavily influenced by its history, particularly the atrocities committed during the Nazi era. This has led to a heightened sensitivity to issues related to human life and ethics. The memory of past abuses in medical research and experimentation by German scientists in the concentration camps during World War II has greatly contributed to a cautious approach in these areas.

2) <u>Ethical Concerns</u>: The German constitution and various other German laws now emphasize the protection of human dignity and the rights of individuals, and human embryos are viewed as possessing a moral status that should be protected. There is a strong ethical stance against the creation and manipulation of embryos for research purposes, as well as concerns about the potential for commodification and the reduction of human life to mere research material.

3) <u>Legal Framework</u>: Germany's "Embryo Protection Act" (1) enacted in 1990, establishes stringent regulations concerning the handling of embryos. This law prohibits the creation of embryos for research and experimentation; it prohibits the creation of not more than three embryos – which must be implanted - as a fertility treatment; and it also prohibits the cryopreservation of embryos unless it is a true emergency within a given fertility treatment. The primary aim is to ensure that embryos are not used in ways that could be seen as exploitative or unethical.

4) <u>Public and Political Opinion</u>: There is significant public and political consensus in Germany against the use of embryos for research or their cryostorage beyond the immediate needs of fertility treatments. This consensus reflects a broader societal commitment to upholding stringent ethical standards in the realm of reproductive technologies.

Italy

Italy's approach to the cryopreservation and experimentation on embryos is governed by Law 40/2004, also known as the "Medically Assisted Reproduction Law" (2). This law reflects Italy's strong stance on bioethical issues that are related to reproductive technologies and the protection of every human embryo's life and dignity.

1) <u>Ethical and Moral Considerations</u>: Italy, with its strong Catholic influence, places a high value on protecting human life from conception. The Roman Catholic Church's teachings significantly impact Italian law and public opinion regarding reproductive technologies and embryonic experimentation. The Church's position that human life begins at conception heavily influences the legal framework aimed at safeguarding embryos.

2) <u>Protection of Embryos</u>: In Italy, there is a broad ethical concern that human embryos should be protected and not subjected to experimentation, cryopreservation, or eugenics. This ethical perspective is closely tied to the deep Christian cultural and religious values in Italian society.

3) <u>Restrictions on Cryopreservation</u>: Law 40/2004 initially prohibited the cryopreservation of embryos except under very specific circumstances related to fertility treatments. This restriction was aimed at preventing the cryostorage of embryos that might be used for purposes beyond medically assisted reproductive needs and to reduce the risk of embryos being discarded or used inappropriately.

4) <u>Limitations on Research</u>: Italy's law prohibits experimentation on human embryos, reflecting a commitment to avoid using embryos as subjects for purely experimental purposes and/or research intended to develop new reproductive technologies.

5) <u>Regulation of Assisted Reproductive Technologies</u>: Law 40/2004 regulates assisted reproductive technologies by stipulating that only specific methods and procedures are allowed. For instance, the law limits the number of embryos that can be created and transferred to reduce and even eliminate the number of surplus cryopreserved embryos.

<u>Louisiana</u>

In 1986, Louisiana passed legislation (3) focused on the protection of frozen embryos, which was part of a broader set of legal measures designed to address ethical and legal concerns related to in vitro fertilization (IVF) and reproductive technologies. Here is a brief overview of Louisiana's regulations that recognize and protect IVF embryos as "juridical persons" and also allows embryo adoption as an option for any surplus IVF embryos.

1) <u>Legal Status of Embryos</u>: Louisiana's legislation recognized the embryos created through IVF as having a certain legal status that warrants protection. The law was designed to ensure that these embryos are treated with respect and care, reflecting the belief in the intrinsic value of human life from its earliest stages.

2) <u>Protection from Death and Harm</u>: The 1986 legislation in Louisiana established guidelines to safeguard frozen embryos from being subjected to unnecessary harm or death. This includes regulations regarding the handling, storage, and disposition of IVF embryos to ensure they are not discarded improperly or used in ways that could be deemed unethical.

3) <u>Informed Consent</u>: The law emphasizes the importance of informed consent for the handling and disposition of embryos. Individuals undergoing IVF treatments are required to be fully informed about the options for their embryos, including the potential risks and procedures involved in their freezing, storage, and future use.

4) <u>Embryo Adoption</u>: Louisiana's legislation was notably forward-thinking in its early recommendation of embryo adoption as an option for surplus frozen embryos. This was a proactive measure aimed at addressing the issue of frozen IVF embryos when the original parents are no longer willing or able to care for them, but who still have the potential for life.

5) <u>Ethical Influence on Policy</u>: Louisiana's early recommendation for embryo adoption reflected a broader ethical stance that sought to honor the potential life of IVF embryos rather than viewing them as mere by-products of fertility treatments. By promoting adoption, the law aimed to provide a compassionate solution to the dilemma of surplus frozen human embryos.

Georgia

Georgia's pioneering Embryo Adoption Law (4), signed into effect in 1997, represents a significant development in reproductive technology law and bioethics. This legislation is notable for several reasons, particularly its focus on ethical public policy and the fundamental legal principles it embodies. Here is a brief explanation of its importance and why it could serve as a model for other states and countries:

1) <u>Respect for Human Life</u>: Georgia's Embryo Adoption Law underscores a commitment to respecting and protecting human life from the earliest stages of development. By recognizing the inherent potential for embryos to develop into human beings and by providing an avenue for their adoption, the law reflects a deep respect for the intrinsic value of human life.

2) <u>Ethical Management of Surplus Embryos</u>: The law addresses the ethical challenges posed by surplus embryos—those created during in vitro fertilization (IVF) procedures that are not used by the original parents. By promoting embryo adoption as a solution, the legislation ensures that these embryos are given a chance at life rather than being discarded or destroyed.

3) <u>Alternative to Destruction</u>: By providing a legal framework for embryo adoption, Georgia's law offers a compassionate alternative to the destruction of surplus frozen embryos. This approach reflects an ethical stance guided by moral considerations that protect human life.

4) <u>Human Dignity</u>: The law protects the principle of human dignity by recognizing that human embryos are human beings who must be treated with respect and care. This aligns with broader legal and ethical standards that emphasize the importance of valuing human life at all stages.

5) <u>Family Law Integration</u>: The law integrates embryo adoption into the family law framework, providing a structured process for adoption that includes legal considerations such as parental rights and responsibilities. This helps ensure that embryo adoption is managed in a manner consistent with existing family law principles that include the longstanding option of adoption.

Georgia's Embryo Adoption Law is truly significant for its ethical approach to handling surplus embryos and for upholding fundamental legal principles such as human dignity from the moment of conception. It is model legislation that truly promotes an ethical solution to the growing crisis of frozen embryos that has been caused by unregulated IVF practices in the US.

NOTE: These links provide access to the various laws discussed in the article above.

1. Germany: https://www.rki.de/SharedDocs/Gesetzestexte/Embryonenschutzgesetz_englisch.pdf?blob=publicationFile
2. Italy: https://pubmed.ncbi.nlm.nih.gov/15333237
3. Louisiana: https://biotech.law.lsu.edu/cases/la/health/embryors.htm
4. Georgia: https://catholicexchange.com/georgia-law-first-in-nation-to-govern-embryo-adoption/

Essay VII

Journal of Bioethics in Law & Culture
Spring 2025 ● vol. 8 issue 1

The Principle of Fraternal Charity, Organ Donation and Embryo Adoption: From Magisterial Condemnation to Magisterial Commendation

Elizabeth Bothamley Rex

https://www.societyofstsebastian.org/spring2025-bioethics-law-culture

The Principle of Fraternal Charity, Organ Donation and Embryo Adoption: From Magisterial Condemnation to Magisterial Commendation

As those of us who are passionate about resolving the plight of hundreds of thousands of frozen embryos continue to faithfully discuss and debate the various difficult and complex issues that are involved, we are actually helping each other to research and rediscover important moral and ethical principles that the Magisterium has so carefully developed and protected over the centuries to preserve the depositum fidei from any error, especially in recent years given the emergence of so many bioethical and moral challenges.

Since the promulgation of Donum vitae in 1987, followed by the promulgation of Dignitas personae in 2008, faithful Catholic scholars have been philosophically and theologically wrestling with each other over the morality or immorality of embryo adoption, a serious question that has been left unanswered over the past 30+ years by the Magisterium of the Catholic Church. During these scholarly debates, I began to wonder why the donation of a human organ or human tissue to a complete stranger is considered not only licit and moral but highly commendable and virtuous, while the donation of a "leftover" human embryo to a complete stranger is considered by some moral theologians to be not only illicit and immoral, but intrinsically "evil."

This discrepancy led me to research the medical history and the theological development regarding the magisterial morality of human organ and tissue donation. This research has been a rewarding and highly illuminating journey. What it uncovered was that moral theologians, and even several Popes themselves, wrestled for decades, both in public and in private, vigorously debating the many complex and difficult pros and cons regarding organ and tissue transplantations until "it became clear that the classical theological treatise about mutilation needed to be revised in light of the scientific achievements in transplantation." (1)

The Magisterium of the Catholic Church faithfully reformed its own moral teachings regarding human organ and human tissue donation from what had previously been considered immoral "mutilation" to what it now considers to be a heroic work of mercy. It took decades of theological debate before the Church concluded that human organ and tissue donation is, in fact, an act of heroic, self-giving love. How did all this happen?

This essay will attempt to provide a brief synopsis of the key individuals who, despite fierce doctrinal and moral opposition from prominent ecclesiastical authorities, were inspired to reconcile and merge the new, promising, life-saving medical procedures involving organ transplantation with Jesus' New Commandment to "love your neighbor as yourself." (2) Their efforts led to the development of what is now heralded as the Principle of Fraternal Charity which has allowed the Magisterium of the Catholic Church to bless and praise the now-well-established morality regarding the heroic donation of blood, tissue and even organs. Perhaps, someday, this very same Principle of Fraternal Charity will also allow the Catholic Church to publicly bless and praise of those who support embryo donation and embryo adoption as further heroic examples of the great commandment to Love our neighbor and by giving the great gift of Life to the very least among us: literally hundreds of thousands of innocent, abused, abandoned, and endangered frozen human embryos.

The Morality of Organ Transplantation Today

I began my research by reviewing the final chapter of Catholic Bioethics and the Gift of Human Life by William E. May, specifically the last section of the chapter dedicated to discussing "Organ Transplants From the Living (Inter Vivos)." The first two sentences were powerful statements explaining what is at stake and why organ and tissue donation is so praiseworthy:

> "Today the transplanting of vital organs, such as a kidney, a portion of the liver, etc., from one living person to another in desperate need of a vital organ is commonplace. We intuitively and instinctively judge that the giving of a part of one's own body to help a gravely or even mortally ill fellow human person is not only morally justifiable but an act of heroic charity." (3)

St. John Paul II, in his *Address to the First International Congress of the Society for Organ Sharing*, said:

"[A] transplant, and even a simple blood transfusion, is not like other operations. It must not be separated from the donor's act of self-giving, from the love that gives life. The physician should always be conscious of the particular nobility of his work; he becomes the mediator of something especially significant, the gift of self which one person has made… so that another might live." (4)

The Catechism of the Catholic Church confirms the morality of organ transplantation, saying:

"Organ transplants are in conformity with the moral law if the physical and psychological dangers and risks to the donor are proportionate to the good that is sought for the recipient." (5)

The Bishops of the United States also fully concur:

"The transplantation of organs from living donors is morally permissible when such a donation will not sacrifice or seriously impair any essential bodily function and the anticipated benefit to the recipient is proportionate to the harm done to the donor." (6)

This said, what followed next was both unexpected and perplexing: Dr. May proceeded to state that while the Magisterium praises the self-giving of vital organs by living persons, nevertheless, there is no single, clear justification for organ donations and it "is still a matter of debate among Catholic theologians." (7)

So, how did this "debate" begin? What was the original controversy, and how did it "evolve" over the years?

The Original "Unanimous" Condemnation of Organ Transplantation as Immoral Mutilation

It is difficult today to imagine that organ transplantation was almost unanimously condemned by moral theologians following the first successful kidney transplant between two twin brothers in Boston back in 1954, but it was, according to Dr. Albert R. Jonsen, Professor Emeritus of Ethics in Medicine at the University of Washington, who in March 2005 delivered a lecture entitled, "From Mutilation to Donation: The evolution of Catholic moral theology regarding organ transplantation" for the Lane Center for Catholic Studies and Social Thought which was published three years later in the Spring 2008 issue of *Urbi et Orbi*.

The purpose of Dr. Jonsen's speech was,

> "to trace the evolution of a particular teaching in Catholic moral theology, namely, the moral permissibility of taking a vital organ for transplantation from one person to another. The history of this teaching reveals a movement from one moral stance, condemnation, to another, commendation. It reveals a move from an individualistic to a social view of the problem and, finally, it shows an internal Catholic moral teaching that had a significant impact on secular moral judgment about the issue." (8)

Prior to the 1950's, when organ transplantation was still medically impossible, Dr. Jonsen explained that the only "mutilation" that was permissible under the "principle of totality" was the removal of a crushed or partially severed limb as the only means to save life. (9) St. Thomas Aquinas touched briefly upon this "principle of totality" which condemned suicide and self mutilation as violations of the Fifth Commandment of the Decalogue, "Thou shalt not kill," as well as violations of God's absolute dominion over the human body. (10) According to the "principle of totality," humans were allowed a "delegated dominion" over their bodies, but only to preserve their bodies in health and life, stating: "any bodily mutilation was justifiable morally if and only if it contributed to the good of the whole body." (11)

Rev. Bert Cunningham's "Bold" Dissertation:
The Morality of Organic Transplantation

An extraordinary exception to this nearly "unanimous" condemnation of organ transplantation by prominent moral theologians was Rev. Bert Cunningham, C.M., A.M., S.T.L. In 1944, a full ten years prior to the first successful transplantation of a single kidney, Fr. Cunningham "wrote a bold doctoral thesis, contesting the judgment of his elders," (12) and submitted his inspiring and illuminating dissertation to the Faculty of the School of Sacred Theology of the Catholic University of America.

> "He drew on a doctrine of Catholic theology that was, at that time, very much discussed: the doctrine of the Mystical Body of Christ. This doctrine proposed that some scriptural references,

largely from the Apostle Paul, suggested that the church could be conceived as an organic body, with Christ as the head and all Christians as members. Cunningham drew moral implications from this theological doctrine. He wrote, 'there exists an ordination of men to one another and as a consequence, an order of their members to one another…. Thus, we contend that men are ordinated to society as parts to a whole and, as such, are in some way ordinated to one another.' This spiritual ordering allows any person to mutilate himself physically for the good of another part of the mystical body (unless the mutilation caused sterilization or great bodily harm)." (13)

Dr. Jonsen continued, emphasizing the extension of this doctrine to all humans, not just Christians:

"Crucial to this theological doctrine is the concept that this body is 'mystical' in the sense that it is not coincident with the visible church: *it contains all humans, even those who do not know that they are part of it, because all humans have been redeemed by Christ. Thus transplantation is morally legitimate between all humans.*" (14) (Emphasis added)

The Importance of Context Regarding Pope Pius XII's Opposition to Cunningham's Thesis

Eight years later, in 1952, Pope Pius XII, delivered a speech to a convention of histopathologists and, while not mentioning Cunningham by name, took direct issue with his doctoral thesis, calling the justification of experiments on individuals for the good of society a distortion of the moral notion of community:

> "Community exists to facilitate exchange of mutual need and to aid each man to develop his personality in accord with his individual and social abilities. Community is not a physical unity subsisting in itself and its individual members are not integral parts of it." (15)

Four years later, on the topic of corneal transplants in an address to a group of ophthalmologists in 1956, Pope Pius XII once again, without citing Cunningham directly, calls into question his 1944 doctoral thesis:

> "We must note a remark that leads to confusion and which we must rectify ... that individuals could be considered parts and members of the whole organism that constitutes 'humanity' in the same manner – or almost – as they are parts of the individual organism of a man. This is inaccurate. Integrity means the bodily unity of a physical organism in which parts have no independent function except in relation to a whole ... in 'humanity' each individual is a value in himself, although related to others." (16)

"Both of these [papal] allocutions," Dr. Jonsen remarks, "emphasize an individualistic rather than a communitarian interpretation of the principle of totality." (17)

While most of the leading moral theologians during this early period cite these statements as authoritative and magisterial censures of organ transplantation, according to Dr. Jonsen, they failed "to set the papal remarks in full context" (18), especially in the important context of Pope Pius XI's 1930 Encyclical that addressed mutilation for the purpose of eugenic sterilization. Dr. Jonsen explains:

> "The two allocutions of Pius XII, while endorsing an individualistic interpretation of the principle of totality and condemning its extension to society, were given at a time when the Pope (and the rest of the world) were deeply concerned about totalitarianism, the political ideology that subordinates individuals to the state. His [Pius XII's] remarks about the principle of totality in relation to experimentation explicitly has the Nazi medical experiments in mind." (19)

Rev. Gerald Kelly, S.J. and the Revision of the Treatise on Mutilation

In 1954, following the first successful kidney transplant between twin brother by Drs. Joseph Murray and John Merrill in Boston's Peter Bent Brigham Hospital, the renowned American Jesuit moral theologian, Fr. John Connery, commented on Fr. Cunningham's thesis in an article that was published in the theological journal, Theological Studies, "Personally, I am in favor of it." Furthermore, following a risky surgery on a pregnant woman in order to save her fetus, he supported the position of undertaking the risk "to sacrifice an organ for the good of another." (20)

The major revision of the treatise on mutilation, however, is primarily attributed to Rev. Gerald Kelly, a preeminent Jesuit moralist, who in 1956 wrote an entire article entitled "Pope Pius XII and the Principle of Totality" in which he made the case "that the Pope's condemnations of mutilation are in their context intended as condemnations of eugenic sterilization and of human experimentation under totalitarian coercion. They were not directly relevant to organ transplantation." (21)

According to Dr. Jonsen, one of Fr. Kelly's major contributions to organ transplantation was to propose that

> "Transplantation is justified by the law of charity, calling on persons to make sacrifices for the good of others, just as Christ had sacrificed himself for the salvation of the world. St. Thomas had affirmed that a person may even give his life for the good of another person." (22)

In his own words, Fr. Kelly stated in one of his articles in Theological Studies that, "Aquinas showed that in giving one's life for his neighbor, one prefers his own good of a higher order … namely, not a physical good but the good of virtue." (23)

Fr. Kelly wrote a second article in 1956, entitled, "The morality of mutilation: toward a revision of the treatise." According to Dr. Jonsen, in this article, Fr. Kelly makes the strong case that:

> "the classical theological treatise about mutilation needed to be revised in light of the scientific achievements in transplantation. Not only could a kidney be taken from a healthy person with relatively low risk, that same organ was very likely to save the life of another. The classical arguments about mutilation, including the papal statements, were valid in their contexts but were inadequate to deal with this new phenomenon (emphasis added). A new formulation was required. He suggested that formulation: 'The rule of morality should be stated: ordinarily, direct self mutilation is permitted only for one's own direct good but, in exceptional circumstances, the law of charity allows it for the benefit of the neighbor (emphasis added).'" (24)

In concluding this important article, Fr. Kelly made a truly remarkable statement about the controversy that had surrounded mutilation and organ donation and which I believe could be very useful during the current controversy that has surrounded embryo transfer and embryo adoption. Dr. Jonsen made the very astute and important observation that while many of Fr. Kelly's contemporary scholars were calling for an explicit papal statement to resolve once and for all the ongoing debate over mutilation and organ transplantation,

> "[Fr. Kelly] felt that the controversy itself was valuable. 'We are learning much from the controversy and we will still learn more, and surely no harm will come from it if moralists avoid the moral errors at which papal statements have been leveled.' Fr. Kelly was a strong advocate of respect for papal teaching but, at

the same time, as a skilled theologian, he recognized that any papal statement called for careful interpretation in the light of context and circumstances." (25)

The Law to Love Our Neighbor and
The Principle of Fraternal Charity

Fr. Kelly's article on the need for a "revision of the treatise" in light of the scientific achievements regarding the life-saving transplantation procedures was widely embraced and succeeded in transforming what was once a condemnation of mutilation and organ transplantation into what is now a commendable and virtuous act of love for one's neighbor. This "new" treatise is often referred to as the "Principle of Fraternal Charity." One of the Catholic Church's highly respected moralists, Bernard Haring, agreed and wrote the following statement in his book entitled, The Law of Christ: "In transplantation, the organ is not destroyed but loving transferred to one's neighbor in order to overcome a hazard to his life." (26)

Conclusion: The Greatest Commandment of All

We must always remember Our Lord's words when forming our conscience and in our search for truth:

> "But when the Pharisees heard that he had silenced the Saducees, they came together. And one of them, a lawyer, asked him a question, to test him. 'Teacher, which is the great commandment in the law?' And he said to him, 'You shall love the Lord your God with all your heart, with all your soul, and with all your

mind. This is the great and first commandment. And a second is like it. You shall love your neighbor as yourself. On these two commandments depend all the law and the prophets.'" (27)

Dr. Jonsen concludes his remarkable lecture on the development of the morality of organ transplantation with the following statement about "giving the gift of life:"

"The law of charity, so central to Catholic morality took its place as the most basic justification of the previously condemned mutilation. In the secular world of medical ethics and law, the non-theological counterpart of charity became the key concept behind the morality of transplantation. That concept was the giving of a gift, donation: and the gift that was given was the "gift of life." (28)

Life is the greatest gift of all, and in concluding this essay, I would like to offer one final quote from a Zenit article dated August 30, 2003, in which two bioethics experts, Dr. Monica Lopez Barahona and Father Ramon Lucas, were interviewed regarding the licitness and morality of embryo adoption.

One of the questions they were specifically asked recapitulated many of the basic concerns previously expressed regarding the morality of organ transplantation. The question posed to them about embryo adoption was this:

Q: "If human life is an absolute and incommensurable value, and if it is necessary to do everything possible to save a person's life, would not the values, recognized by a personalist and Christian anthropology, remain subordinated: the right of the child to be gestated in the womb of his mother; the right of the child to be born in a context that also guarantees the balanced growth of the personality; the value of maternity as a personal event which excludes as a line of principle the separation of the biological, physiological and emotional processes; the representation of human procreation as an interpersonal act of a triadic nature – father, mother, child?

A: "As said at the beginning, the difference must be made again, between moral acts and rights.

"All those values recognized by a personalist and Christian anthropology are values that are arranged in order of importance according to a value that is original and prior to them all, as is the assumption and necessary condition for them to be present: [That value is] human life.

"Human Life has priority over these enunciated values. Otherwise, it would have to be concluded that the existence of those who have seen those rights violated has no meaning – and this is absurd, among other things, because it would exclude a good part of humanity.

"Moreover, the objection is untenable as, in fact, the frozen embryo has already lost those rights: its biological mother has abandoned it; so has its family; the biological maternity has been completely perverted and subverted; the same happened with the act of human procreation. …

The evil is already done. Only prenatal adoption can, to a very limited degree, try to repair in some way such injustices." (29)

The primacy of life and the New Commandment to love our neighbor, as developed in the Principle of Fraternal Charity, are two powerful arguments that defend both organ transplantation and embryo adoption. May we ponder deeply the primacy of the Gift of Life and the Law of Love as we continue to seek the truth, and to proclaim the truth in love, "veritatem facientes in caritate." (30) Countless frozen embryos are not only our neighbors; they are also the least of our brethren, and what we do to them we do to Jesus.

Today the Magisterium of the Catholic Church commends organ and tissue transplantation inter vivos as a corporal work of mercy and an example of heroic virtue. Likewise, embryo adoption inter vivos is clearly also a work of mercy and, in fact, encompasses all of the spiritual and corporal works of mercy for the very least of our brethren. Jesus, Perfect God and Perfect Man, was once a one-cell zygote in the womb of His Mother. Clearly, what we do for tiny embryos, we do unto Jesus Himself. Let us pray that the Magisterium will soon proclaim that the Principle of Fraternal Charity equally applies to the least of our brethren: frozen embryos.

Endnotes

1) Albert R. Jonsen, Ph.D., *Urbi et Orbi*, Spring 2008, 5.
2) St. Matthew, 22:39.
3) William E. May, *Catholic Bioethics and the Gift of Human Life* (Huntington, IN: Our Sunday Visitor Publishing Division, 2008), 353.
4) Pope John Paul II, Address to the First International Congress of the Society for Organ Sharing, *L'Osservatore Romano*, English ed. (June 24, 1991), 2.
5) *Catechism of the Catholic Church*, Second Edition (Libreria Editrice Vaticana, 2007), no. 2296.
6) National Conference of Catholic Bishops, Ethical and Religious Directives for Catholic Health Care Services (Washington, DC: United States Catholic Conference, 1995), directive no. 30.
7) May, op cit., 354.
8) Jonsen, op. cit., 4.
9) Ibid., 4.
10) Summa Theologica, II-II, 65,1.
11) Ibid., 4.
12) Ibid., 4.
13) Ibid., 4.
14) Ibid., 5.
15) Acta Apostolici Sedis 1952; 44: 786.
16) Acta Apostolici Sedis 1956; 48: 446.
17) Jonsen, op.cit., 5.
18) Ibid., 5.

19) Ibid.
20) Notes on Moral Theology, Theological Studies, 1954; 15.
21) Jonsen, op. cit., 5.
22) Ibid., 5, 8.
23) Ibid.
24) Theological Studies, 1956; 17: 342.
25) Jonsen, op. cit., 8.
26) Bernard Haring, The Law of Christ, III, 242.
27) St. Matthew, 22:34-40.
28) Jonsen, op.cit., 8.
29) Monica Lopez Barahona and Rev. Ramon Lucas, Zenit, August 30, 2003: "Why Adoption of Frozen Human Embryos Could Be Acceptable;" http://www.zenit.org/en/articles/why-adoption-of-frozen-human-embryos-could-be-acceptable
30) Ephesians 4:15.

The Zygote of Christ
& The Mystery of Man

Selected Essays
Written By
Francis Etheredge & Elizabeth Rex

Front Cover Art Commentary
The Zygote of Christ
Nellie Edwards

The Story Behind The Zygote of Christ
Nellie Edwards

God's grace is truly a wonder-full gift! As a young girl, I had a deep desire to do fine art but had only a thimble of talent. I tried many times throughout my life to develop what little ability I had, but my attempts always proved that I would need to one day take lessons!

I married at the age of twenty, and before long, our growing family made art lessons seem impractical. I knew I would have to wait until the kids were grown. I could not have imagined what God would do for me to realize my dream!

The Nudge: In 2007, after an act of obedience to God, which I knew would cause our family's income to shrink, something unexpected happened. As a result of this obedience, I found myself with a PC Tablet, and within a short time, had a sudden prompting to create a portrait of Blessed Kateri Tekakwitha, the "Lily of the Mohawk." I resisted the idea because of my lack of ability, but the idea kept coming, so I gave in!

The PC Tablet and stylus, with which I had no prior experience, served as my canvas and brush, and after seven months of work, I was amazed at the result! God's grace allowed it to appear as though done by an accomplished artist, not by someone with no formal training! Eventually, the Knights of Columbus requested *Holding on to Faith* for the cover of the canonization issue of *Columbia* magazine. Other publishers followed, including Our Sunday Visitor, for a biography of Kateri, which I found at the Vatican bookstore when in Rome for her canonization!

Mother of Life: In 2011, while in the early stages of a new depiction of the Blessed Mother, a very strong awareness came to me that this *had to be Our Lady of Guadalupe*! I was taken aback since I always had great reverence for the miraculous image which God left on the tilma of St. Juan Diego! I truly felt it would be irreverent to attempt such! But the prompting was unrelenting. After praying, I felt real peace about it and so began.

Because Mary's face was already in the profile aspect, I realized I had the opportunity to show the full-termness of Unborn Jesus, the *"Light of the World,"* glowing in His mother's womb! I also discerned that God wanted the Blessed Mother to be kneeling in Adoration of her unborn Savior Son; this, to encourage all believers to likewise spend time in *Adoration*! In nine months, it was finished and I titled it *"Mother of Life."* This, too, was requested for the cover of *Columbia* magazine and, eventually, by many others. Soon, the beloved theologian, Dr. Scott Hahn, posted it on social media, which propelled it around the world! What amazed me, too, was the more than 800,000 who responded to a post of *Mother of Life* by a priest from Mexico!

To me, this was the result of *Divine Intervention,* and I realized how vital it was that I had obeyed God, as previously mentioned, in order to receive the talent to create visual aids for God's good purpose! I am deeply grateful for the reports I continue to receive from people around the country who tell me that *Mother of Life* has moved abortion-bound women to choose life! Again, this is only due to the grace of God. It is the Lord who guides the brush, but, again, He tested me to see if I would obey what would mean a reduction in our family's income.

How The Zygote of Christ came about: In 2012, my husband was being treated for cancer. A day came when I had to get him to the hospital as he was in quite some pain. After running tests, the oncologist told me that there was nothing that could be done for him, which was devastating news! I wondered if I should have him moved to another hospital. First, though, I took up my rosary, which I pray daily, and asked the Blessed Mother for her intercession, saying, "I need you, Mama!" As soon as I finished praying, I did something very unlikely; I turned on the TV! This felt strange since we had gotten rid of our television years earlier, and this certainly did not seem like the time for such a thing!

I was shocked when the set came on, since, without changing channels, there was EWTN's *Women of Grace* show; this, in a non-Catholic hospital! I immediately sensed that Divine Intervention was in motion! Almost instantly, the host, Johnnette Benkovic, asked her three guests, "At what point did you know you had to let your husbands go?" I gasped! The first woman to answer said, "The Lord spoke to my heart and said, 'He's been on loan to you; I'm calling him home!'"

I was truly stunned, to say the least! This was the fastest answer to prayer I had ever experienced! At that moment, I knew the Lord was calling my husband home. And with that knowledge came a peace that only God can give!

A Providential Meeting: A year later, in 2013, I attended a three-day Catholic Marketing Network conference in New Jersey, where my distributor was debuting my artwork. On the first day, a woman came by the exhibit and told me how much she appreciated the way I depicted the *Word Made Flesh*. She seemed familiar, but I didn't dwell on it. She returned the second day to visit on the subject of the *Incarnation of Christ,* and I wondered if we had met before but shrugged off the idea. On the third and final day, she said to me, "Nellie, I think you and I should go on the *Women of Grace* show together." Suddenly, I knew why she seemed so familiar! With tears flowing, I said, "You're the one whose words brought me such peace during my husband's final days!" We embraced each other, crying tears of joy and amazement at how God had brought us together!

Front Cover Art Commentary by Nellie Edwards 171

We later realized that God had even given us the same initials: E.M.E. - Ellen Marie Edmonds and E.M.E. for Eleanor Mary Edwards! (Nellie is my nickname) We both felt that God was going to accomplish something with us!

Help me Lord: A few weeks later, Ellen Marie called and suggested I pray about doing *The Zygote of Christ*! At first, I thought it would be impossible! How could I create a composition of something almost invisible to the naked eye - a single-cell zygote? But I soon realized this must be God's will because of the way He had placed Ellen Marie in my life! Looking up, I prayed, "Lord, I know You want this, but I need help... a lot of help!"

Almost immediately, the words of 1 Kings 8:27 came to mind, which says, *"The heavens cannot contain the glory of God."* I knew then that the composition needed to look cosmic, since the greatest cosmic event of all time was the moment that God became man!

I began right away, with the zygote in the center (magnified thousands of times). Through prayer, I was guided to include God the Father, looking down in approval, and the Holy Spirit at the moment of the Incarnation! I was also led to show an abbreviation of the fetal development of Jesus, at twenty-one days, when His tiny Sacred Heart would first beat; at five months gestation; and finally, as the newborn Messiah! I also discerned to include the faces of the Risen Christ and Mary, His Mother, subtly, within the zygote itself!

When the painting was complete, I realized I had done something without realizing its meaning; the outer edge initially formed a perfect circle, and I knew it didn't look right. I felt led to make random indentations. I was overwhelmed, to realize later, that this represents the *monstrance*, housing the *Real Presence* of Jesus, even at zygote stage!!

When the work was finally finished, I knew the image had a mission: to help people understand the sanctity of every human life, from the very first moment of conception! Again, our God Made Man began as a zygote, and millions of our zygote-stage brothers and sisters are frozen and doomed to die without an adoption option! God loves them no less than any of his born children. We ought to treat the "least of His little ones" accordingly!

My heartfelt gratitude to Elizabeth Rex, and all who are working with her, for this great cause! Let's pray their efforts will ultimately save countless human lives.

www.ingramcontent.com/pod-product-compliance
Lightning Source LLC
LaVergne TN
LVHW052100090426
835512LV00036B/2856